WRITING FOR COMPUTER SCIENCE

Springer
Singapore
Berlin
Heidelberg
New York
Barcelona
Budapest
Hong Kong
London
Milan
Paris
Tokyo

Justin Zobel

WRITING for COMPUTER science

THE ART OF EFFECTIVE COMMUNICATION

Justin Zobel
Multimedia Database Systems Group
RMIT Research
732 Swanston Street
Carlton 3053
Melbourne
Australia

Library of Congress Cataloging-in-Publication Data

Zobel, Justin, 1963 –
Writing for Computer Science : the art of effective communication / Justin Zobel
p. cm.
Includes bibliographical references and index.
ISBN 9813083220
1. Technical writing. 2. Communication of technical information. I. Title.
T11.Z62 1997 97-13754
808'.066004--dc21 CIP

ISBN 981-3083-22-0

Reprinted 1998

Typesetting: Camera-ready by author
Printed in Singapore
SPIN 10699217 5 4 3 2 1

in memory of my mother

Contents

Preface

This writing seemeth to me ... not much better than the noise or sound which musicians make while they are in tuning their instruments.

Francis Bacon
The Advancement of Learning

No tale is so good that it can't be spoiled in the telling.

Proverb

A scientific article is a description of new ideas and a demonstration of their correctness. An article can remain relevant for a remarkably long time and, if published in a major journal, may be read by thousands of other scientists.

Unfortunately many scientists do not write well. Bacon's comment was made four hundred years ago, but applies to much science writing today. Indeed, perhaps we should not always expect scientists to write well—the skills required for science and writing are rather different.

But that does not mean that scientists should be content to write badly. Every scientist whose work is affected by a poorly-written paper will suffer: ambiguity will lead to misunderstanding; omissions will frustrate; obscurity will make readers struggle to reconstruct the author's intention. Effort used to understand the form of an article—its structure and syntax—is effort not used to understand its content. And no tale is so good that it can't be spoiled in the telling. Irrespective of the importance and validity of its argument, a report will not convince anybody of anything if it is difficult to understand. The more important the results (or the greater their surprise value) the better the supporting arguments and their presentation should be.

Publication not only makes knowledge available, it establishes the authors as the creators of that knowledge. Authorship has obvious rewards such as position and promotion, and has other rewards too; for example, by and large it is the basis on which the scientific Nobel prizes are awarded. But authorship implies responsibility. Public mistakes not only make a scientist look foolish, they can hurt a career.

Moreover, writing is not just a means of making ideas public. Another important aspect of it is that the discipline of stating ideas as organized text forces authors to formulate and codify their thoughts. Vague concepts become concrete; the act of writing suggests new concepts to consider; and written material is easier to discuss with colleagues. That is, writing is not the end of the research process—it is integral to it. Only the styling of a paper, the polishing process, truly follows the research.

Taking another perspective, scientific papers are a way of communicating ideas, of copying thoughts between minds. Communication is at its most effective when the medium is as free as possible from distortion, which in this case is bad writing of any kind. Such distortion can be reduced by writing with clarity and simplicity, and by making effective use of stylistic conventions.

About this book

"Writing for Computer Science" is an introduction to the style and presentation of scientific writing with computing or mathematical content, and is intended for researchers and senior students. For the most part it consists of suggestions describing what I believe to be good style; some of these suggestions are accepted wisdom, some are controversial, and some are my opinions. I have concentrated on writing of articles, but the same skills apply to writing of theses and texts, and much of the material is applicable to general technical and professional communication. Although this book is brief, it is designed to be comprehensive. Some readers may be interested in exploring topics further, but for most readers this book will I hope be sufficient.

Style is to some degree a matter of taste. The advice in this book should not be treated as a code of law to be rigidly obeyed: it is my opinion, and there are inevitably situations in which the "correct" style simply seems wrong. But generally there are good reasons for writing in a certain way. Almost certainly you will disagree with some of the points given in this book, but at least exposure to another's opinion should lead you to justify your own choice of style, rather than by habit continue with what may be poor writing. A good principle is: by all means break a rule, but have a good reason for doing so.

This book is about writing, but is not just about text. It includes material on the major facets of writing in computer science: design of papers (Chapter 1), writing style (Chapters 2 to 4), mathematical style (Chapter 5), design of figures and graphs (Chapter 6), presentation of algorithms (Chapter 7), development of hypotheses and conduct of experiments (Chapter 8), editing, including a checklist to be used for revision of papers (Chapter 9), refereeing (Chapter 10), and presentation of talks (Chapter 11). There is also a series of exercises to help develop writing skills, and a short bibliography of style guides and other texts.

This book has been written with the intention that it be browsed. Readers should not memorize it or learn it as a dry set of rules. Read it, absorb whatever advice seems of value to you, find out where the reference lists are; then consult it for specific problems.

Many people helped with this book in one way or another, in particular Alistair Moffat, who contributed to Chapters 1, 6, 7, 9, and 10; and Philip Dart, who also contributed to Chapter 10. I am grateful to both Alistair and Philip for our research collaborations and for teaching me much about writing. I am also grateful to Isaac Balbin, Gill Dobbie, Evan Harris, Mary and Werner Pelz, Kotagiri Ramamohanarao, Ron Sacks-Davis, James Thom, Ross Wilkinson, and Hugh Williams; my research students; the students who participated in my research methods lectures; Michael Fuller for his thorough proofreading; Ian Shelley at Springer-Verlag for his support and assistance; and Rodney Topor, who introduced me to many of the topics of good style. Most of all I am grateful to my wife, Penny Tolhurst.

Justin Zobel
March 1997

1 Designing an article

Any human thing supposed to be complete, must for that reason infallibly be faulty.

Herman Melville
Moby Dick

I used to think about my sentences before writing them down; but ... I have found that it saves time to scribble in a vile hand whole pages as quickly as I possibly can ... Sentences thus scribbled down are often better ones than I could have written deliberately.

Charles Darwin
Autobiography

Science is a system for accumulating reliable knowledge. Broadly speaking, the process of science begins with observations; which are developed into hypotheses, tested by proof or experimentation; yielding results that can be described in a paper; which is published after independent reviewing. Each new contribution builds on a bed of existing concepts that are known and trusted. New research may be wrong or misguided, but the process of refereeing eliminates most work of poor quality, while the scientific culture of questioning ideas and requiring convincing demonstrations of their correctness weeds out (perhaps gradually) published falsehoods.

Good writing is a crucial part of the process of accumulating knowledge. For an idea to survive, other scientists must be persuaded of its relevance and correctness—not with rhetoric, but in the established framework of a scientific publication. New ideas must be explained clearly to give them the best possible chance of being understood, believed, remembered, and used. Authors need to:

- Describe the position of the new idea in the body of scientific knowledge.
- Formally state the idea, often as a theory or hypothesis.
- Explain what is new about the idea, or what contribution the paper is making.
- Justify the theory, by methods such as proof or experiment.

A typical article consists of the arguments, evidence, experiments, proofs, and background required to support and explain a central hypothesis. In contrast, the process of research that leads to an article can include dead ends, invalid hypotheses, misconceptions, and experimental mistakes. With few exceptions these do not belong in an article—it should be an objective addition to scientific knowledge, not a description of the path the author took to the result.

Kinds of publication

Scientific results can be published as a book, a thesis, a journal article, a complete article or extended abstract in a conference or workshop proceedings, a technical report, or a manuscript. Each kind of publication has its own characteristics. Books are usually texts that tend not to contain new results and do not to the same degree justify the correctness of the information they present. The main purpose of a text is to collect information and present it in an accessible, readable form, and thus texts are generally better written than are papers.

A thesis is usually a deep exploration of a single problem, often in effect a long article or a series of articles on the same topic. Journals and conference proceedings consist of contributions that range from substantial papers to extended abstracts (which might be better described as abridged papers). A journal article is usually the end product of the research process, a careful presentation of new ideas that has been revised, sometimes over several iterations, according to referees' suggestions and criticisms. In contrast to books—which represent the authors' opinions, unalloyed by the judgement of other scientists—the content of an article must be defended and justified. An article or extended abstract in a conference proceedings can likewise be an end-product, but conferences are also used to report work in progress. Conference articles are usually refereed, but with more limited opportunities for iteration and revision, and may be constrained by strict length limits.

Prior to a paper appearing in a refereed venue it may be available as a manuscript or technical report. These forms of publication have the advantage of making the work available quickly—virtually instantly if the paper or its abstract is available on the world wide web—but, because of the lack of refereeing, readers have less confidence in the work's validity.

Organization

Scientific articles usually follow a standard structural recipe that allows readers to quickly discover the main results, and then, if interested, to examine the supporting evidence. Many readers accept or reject conclusions based on a quick scan, not having time to read all the articles they see. A well-structured article accommodates this behaviour by having important statements as near the beginning as possible.

A typical article has most of the following components.

Title and author

Papers begin with their title and information about authors including name, affiliation, and address. The convention in computer science is to not give your position, title, or qualifications; but whether you give your name as A. B. Cee, Ae Cee, Ae B. Cee, or whatever, is a personal decision. Use the same style for your name on all your papers, so that they are indexed together. Include an e-mail address if you have internet access.

Also include a date. If you are using a word processor, take the trouble to type in the date rather than using "today" facilities that print the date on which the document was last processed, or later you may not be able to tell when the document was actually completed.

Abstract

An abstract is typically a single paragraph of about fifty to two hundred words. The function of an article's abstract is to allow readers to judge whether or not the article is of relevance to them. It should therefore be a concise summary of the aims, scope, and conclusions of the article. There is no space for unnecessary text; an abstract should be kept to as few words as possible while remaining reasonably informative. Irrelevancies, such as minor details or a description of the structure of the paper, are inappropriate, as are acronyms, abbreviations, and mathematics. Sentences such as "we review relevant literature" should be omitted.

The more specific an abstract is, the more interesting it is likely to be. Instead of writing "space requirements can be significantly reduced", write "space requirements can be reduced by 60%"; or instead of writing "we have a new inversion algorithm" write "we have a new inversion algorithm, based on move-to-front lists".

Many scientists browse research papers outside their area of expertise. Authors should not assume that all likely readers will be specialists in the topic of their article—abstracts should be self-contained and written for as broad an audience as possible. Only in rare circumstances will an abstract cite another paper (for example, when one paper consists entirely of analysis of results in another), in which case the reference should be given in full, not as a citation to the bibliography.

Introduction

An introduction can be regarded as an expanded version of the abstract. It should describe the article's topic; the problem being studied; the approach to the solution; the scope and limitations of the solution; the conclusions, with enough detail to allow readers to decide whether or not they need to read further; and, often, the article's structure. It should also include motivation: the introduction should explain why the problem is interesting, what the relevant scientific issues are, and why the solution is a good one. That is, it should explain why the paper is worth reading.

The introduction can discuss the importance or ramifications of the conclusions but should omit supporting evidence, which the interested reader can find in the body of the paper. Relevant literature can be reviewed in the introduction, but complex mathematics belongs elsewhere.

An article isn't a story in which results are kept secret until a surprise ending—the introduction should clearly tell the reader what in the article is new and what the outcomes are. There may still be a little suspense: revealing what the results are does not necessarily reveal how they were achieved. If however the existence of results is concealed until later on, the reader might assume there are no main results and discard the paper as worthless.

Survey

Few results or experiments are entirely new. Most often they are extensions of or corrections to previous research—that is, most results are an incremental addition to existing knowledge. A survey is used to compare the new results to similar results in the literature, and to describe existing knowledge and how it is extended by the new results. A survey can also help a reader who is not expert in the field to understand the article and may point to standard references such as texts or survey articles.

The survey can come early in an article, to describe the context of the work, and might in that case be part of the introduction; or the survey can follow or be part of the main body, at which point a detailed comparison between the old and the new can be made. If the survey is late in an article, it is easier to present the surveyed results in a consistent terminology, even when the original presentations have differing nomenclature and notation.

In many papers the survey material is not gathered into a discrete section, but is discussed where it is used—background material in the introduction, analysis of other researchers' work as new results are introduced, and so on. This approach is often easier on the reader.

Results

The body of an article should present the results: provide necessary background and terminology, explain the chain of reasoning that leads to the conclusions, provide the details of central proofs, summarize any experimental data, and state in detail the conclusions outlined in the introduction. It should also contain careful definitions of the hypothesis and major concepts, even those described informally in the introduction.

The structure should be evident in the section headings. Since the body can be long, narrative flow and a clear logical structure are essential.

The body should be reasonably independent of other papers. If, to understand your paper, the reader must find specialized literature such as your earlier papers or an obscure paper by your supervisor, then its audience will be limited.

Most experiments yield far more data than can be presented in a paper of reasonable length. Important results can be summarized in a graph or a table, and other outcomes reported in a line or two. It is acceptable to state that experiments have yielded a certain outcome without providing details, so long as those experiments do not affect the main conclusions of the paper (and have actually been performed). Similarly, there may be no need to include the details of proofs of lemmas or minor theorems. This does not of course excuse authors from conducting the experiments or proving to themselves that the results are correct, but such information can be kept in logs of the research rather than included in the paper.

Discussion of proofs or experimental results should be part of their presentation. It is hard to read papers in which results or theorems are listed in one section and analyzed elsewhere; a better approach is to analyze the results as they are presented, particularly since experiments or theorems often follow a logical sequence in which the outcome of one dictates the parameters of the next. When describing results, it is helpful—although not always possible—to begin with a brief overview of whatever has been observed, and to use the rest of the discussion for amplification rather than further observations.

There are several common ways for structuring the presentation of the results. Perhaps the most common structure is a chain, in which the results and background dictate a logical order. First might come, say, a problem statement, then a review of previous solutions and their drawbacks, then the new solution, and finally a demonstration that the solution improves on its predecessors.

Other structures may however be preferable for some kinds of results. One option is to structure by specificity, an approach that is particularly appropriate for results that can be divided into several parts. The material is first outlined in general terms, then the details are progressively filled in. Most technical papers have this organization at the high level, but it can also be used within sections.

Another structure is by example, in which the idea or result is initially explained by, say, applying it to some typical problem. Then the idea can be explained more formally, in a framework the example has made concrete and familiar.

The body can also be structured by complexity. For example, a simple case can be given first, then a more complex case can be explained as an extension, thus avoiding the difficulty of explaining basic concepts in a complex framework. This approach is a kind of tutorial: the reader is brought by small steps to the full result. For example, a mathematical result for an object-oriented programming language might initially be applied to some simple case, such as programs in which all objects are of the same class. Then the result could be extended by considering programs with inheritance.

(Structuring by complexity is good for a paper but bad for a research program. It is not uncommon to see a paper in which the authors have solved an easy case of a problem, say optimizations for iteration-free programs, motivated by hopeful claims such as "we expect these results to throw light on optimization of programs with loops and recursion"; but all too often the follow-up paper never appears.)

Summary

The closing summary, or conclusions, is used to draw together the topics discussed in the paper and includes a concise statement of the paper's important results. The summary can also look beyond the current context to other problems that were not addressed, to questions that were not answered, to variations that could be explored; and there may be speculation, such as discussion of possible consequences of the results.

Bibliography

An article's bibliography, or more properly its set of references, is a list of papers, books, and reports cited in the text of the article. No other items should be included.

Appendices

Some articles have appendices giving detail of proofs or experimental results, and, where appropriate, material such as listings of computer programs. The purpose of an appendix is to hold bulky material that would otherwise interfere with the narrative flow of the paper, or material that most readers will not refer to; thus appendices are not usually necessary.

The first draft

For the first draft many writers find it helpful to write freely—without particular regard to style, layout, or even punctuation—so that they can concentrate on presenting a smooth flow of ideas in a logical structure. Worrying about how to phrase each sentence tends to result in text that is clear but doesn't form a continuous whole, and authors who are too critical on the first draft are often unable to write anything at all. If you tend to get stuck, just write anything, no matter how awful; but be sure to delete any ravings later.

A consequence of having a sloppy first draft is that you must edit and revise carefully; initial drafts are often turgid and full of mistakes. But few authors write well on the first draft anyway; the best writing is the result of frequent, thorough revision.

Mathematical content, definitions, and the problem statement should, however, be made precise as early in the writing process as possible. The hypothesis and the results flow from a clear statement of the problem being tackled; describing the problem forces you to consider in depth the scope and nature of the research; and if you find that you cannot describe the problem precisely then perhaps your understanding is lacking or the ideas are insufficiently developed.

The writing should begin before the research is finished. Writing is a stimulus to research, suggesting fresh ideas and clarifying vague concepts and misunderstandings; and developing the presentation of the results will often suggest the form the proofs or experiments should take. Gaps in the research may not be apparent until it has been at least preliminarily described. Research is also a stimulus to writing—fine points are quickly forgotten once the work is complete.

Composition

There are many texts that describe the process of assembling a technical article; see for example Maeve O'Connor's *Writing Successfully in Science* [19] or Bruce M. Cooper's *Writing Technical Reports* [10].

The technique I use for composing is to brainstorm, writing down in point form what I think has been achieved and what the results are. I then prepare a skeleton, choosing results to emphasize and discarding material that on reflection seems irrelevant, and work out a logical sequence of sections that leads the reader naturally to these results. A useful discipline is to choose the section titles before writing any text, because if material to be included doesn't seem to belong in any section then the paper's structure is probably faulty. The introduction is completed first and includes an overview of the article's intended structure, that is, an outline of the order and content of the sections. When the structure is complete, I sketch each section in perhaps 20 to 200 words. This approach has the advantage of making the writing task less daunting—it is broken into parts of manageable size.

When the body and the closing summary are complete, the introduction usually needs substantial revision because the arguments presented in the article mature and evolve as the writing proceeds. The final version of the abstract is the last part to be written.

For a novice writer who doesn't know where to begin, a good starting point is imitation. Choose a paper whose results are of a similar flavour to your own, analyze its organization, and sketch an organization for your results based on the same pattern. The habit of using similar patterns for papers—their standardization—helps to make them easier to read.

Once the paper is designed it must be written with clarity and style, the topic of Chapters 2 to 7. Any experiments should be conducted and described to an appropriate standard, the topic of Chapter 8. And finally the paper must be edited, the topic of Chapter 9.

Choice of word-processor

When you start an article you need to choose a word-processor. The choice is dictated by availability, but also by how well the available word-processors cope with the demands of authoring a technical article. Many articles involve—in addition to text—figures; tables; mathematics; use of multiple fonts and point sizes; and cross-references to figures, tables, equations, sections, and bibliographic entries. These demands can stretch the facilities of even the most sophisticated word-processor.

Further problems are presented by the lifecycle of technical articles. For example, an article might initially be drafted for circulation amongst colleagues, revised for submission to a conference then accepted after further revision and experiments; but because the paper is too long, some text must be omitted. Subsequently, after rethinking, new work, and reintroduction of omitted text, the paper is combined with a report on earlier work and submitted to a journal, where, after revision to meet referees' comments, it is accepted, perhaps three years after the initial draft was written. Word-processors need to be able to handle this high level of revision and re-organization.

There are, broadly speaking, two kinds of word-processor, the visual or WYSIWYG style and the compiler style typified by `troff` and LaTeX, which translate marked-up text into a page description language such as PostScript. The visual word-processors are generally superior at production of documents for immediate use such as letters, and for first drafts, but for science writing the compiler word-processors are preferable.

There are many reasons for this. The compiler word-processors are less susceptible than the visual word-processors to software revision—the latter often use proprietary formats that change at each release, so that authors may have to revise a paper in a different version of the software to that used for the original draft; and the same version of the software must be available to authors jointly working on a paper. In contrast, the compiler word-processors have historically been stable; and the use of plain files for storing text means that co-authoring papers via e-mail and so on is straightforward. Also, the compiler word-processors usually provide some facility for commenting-out text, making omission and re-inclusion straightforward, and have macro facilities that make it easy to generate multiple distinct documents (such as a conference version and a more complete technical report) from one source file.

Perhaps most importantly, documents produced with visual word-processors can look amateurish, particularly if mathematics is involved. Many journals now use LaTeX for typesetting, for good reason—the result is professional. The LaTeX word-processing system was used for this book, and is today the best word-processor for science writing.

2 Writing style: general guidelines

Everything written with vitality expresses that vitality;
there are no dull subjects, only dull minds.

Raymond Chandler
The Simple Art of Murder

It is a golden rule always to use, if possible, a short old Saxon word. Such a sentence as "so purely dependent is the incipient plant on the specific morphological tendency" does not sound to my ears like good mother-English—it wants translating.

Charles Darwin
Letter to John Scott

There are many ways in which an idea can be expressed in English; writing can be verbose or cryptic, flowery or plain, poetic or literal. The manner of expression is the writing style. Style is not about correct use of grammar, but about how well the text communicates with likely readers.

Conventions and styles are valuable because some forms of presentation are difficult to understand or are simply boring, and because conformity to commonly-used styles reduces the effort required from readers.

Flouting an established convention has the impact of this exclamation! It arrests attention and distracts from the message; unless, of course, the message is that a convention is being flouted.

Science writing must by its nature be prosaic—the need for it to be accurate and clear makes poetry inappropriate. But this does not mean that science writing has to be dull. It can still have style, and moreover the desire to communicate clearly is not the only reason to make good use of English. Lively writing suggests a lively mind with interesting ideas to discuss, while poor usage is distracting, suggests disorganized thinking, and prejudices readers against whatever is being presented.

In this chapter, and in chapters 3 and 4, I consider style for text, including issues that are specific to science writing and general issues that many scientists ignore. Most of the points concern the basic aims of science writing: to be unambiguous, clear, simple, correct, interesting, direct. Perhaps the best text on style and clarity for English is William Strunk and E. B. White's wonderfully concise *The Elements of Style* [7]. Also excellent are Ernest Gowers's *The Complete Plain Words* [3], H. W. Fowler's *Modern English Usage* [2], and Eric Partridge's *Usage and Abusage* [6].

Economy

Text should be taut. The length of a paper should reflect its content—it is admirable to say much in a small space. Every sentence should be necessary. Papers are not made more important by padding with long-winded sentences; they are made less readable. In the following example, the italicized text can be discarded without affecting the intent.

> *The volume of information has been rapidly increasing in the past few decades. While computer technology has played a significant role in encouraging the information growth, the latter has also had a great impact on the evolution of computer technology in processing data throughout the years. Historically, many different kinds of databases have been developed to handle information, including the early hierarchical and network models, the relational model, as well as the latest object-oriented and deductive databases. However, no matter how much these databases have improved, they still have their deficiencies.* Much information is *in* textual *format.* This unstructured *style of* data, *in contrast to the old structured record format data,* cannot be managed properly by

> *the* traditional database models. Furthermore, *since so much information is available,* storage and indexing are not the only problems. We need to ensure that relevant information can be obtained upon querying *the database.*

Waffle, such as the italicized material above, is deadwood that the reader must cut away before they can get to the meaning of the text.

Taut writing is a consequence of careful, frequent revision. Aim to delete superfluous words, simplify sentence structure, and establish a logical flow; to convey information without unnecessary dressing. Revise with a critical frame of mind, and absolutely not with a sense of basking in your own cleverness. Be egoless—ready to dislike anything you have previously written. Expect to revise several times, perhaps many times.

If someone dislikes anything you have written, remember that it is readers you need to please, not yourself. Again, it helps to set aside your ego. For example, when making changes to your paper in response to comments from a referee, you may find that the referee makes a claim that is quite wrong. However, rather than telling yourself "the referee is wrong", ask yourself "what did I write that led the referee astray?" Even misguided feedback tells you something about your writing.

Text can be condensed too far. Don't omit words that make the writing easier to understand.

✗ Bit-stream interpretation requires external description of stored structures. Stored descriptions are encoded, not external.

✓ Interpretation of bit-streams requires external information such as descriptions of stored structures. Such descriptions are themselves data, and if stored with the bit-stream become part of it so that further external information is required.

Tone

Science writing should be objective and accurate. Many of the elements that give literature its strength—nuance, ambiguity, metaphor, sensuality—are inappropriate for technical work. In contrast to popular science writing, the primary objective is to inform, not entertain. On the other hand, use of turgid, convoluted language is perhaps the most common fault in scientific writing. A direct, simple, even incisive, style is appropriate: aim for austerity, not pomposity.

Simple writing follows from a few simple rules:

- Have one idea per sentence or paragraph and one topic per section.
- Have a simple, logical organization.
- Use short words.
- Use short sentences with simple structure.
- Keep paragraphs short.
- Avoid buzzwords and clichés.
- Avoid excess, in length or style.
- Omit any unnecessary material.
- Be specific, not vague or abstract.
- Only break one of these rules if there is a good reason to do so.

Sometimes a long word or a complex sentence is the best option; use them when necessary, but not otherwise.

Another common fault in science writing is to overqualify, that is, to modify every claim with caveats and cautions. Such writing is a natural consequence of the scientist's desire to not make unfounded claims, but it can be taken too far.

✗ The results show that, for the given data, less memory is likely to be required by the new structure, depending on the magnitude of the numbers to be stored and the access pattern.

✓ The results show that less memory was required by the new structure. Whether this result holds for other data sets will depend on the magnitude of the numbers and the access pattern, but we expect that the new structure will usually require less memory than the old.

The first version is vague; the author has ventured an opinion that the new structure is likely to be better, but has buried it.

Use direct statements and expressions involving "we" or "I"—that is, the active voice—to make reading more pleasant and to help distinguish new results from old. (Voice is discussed on page 41.) There is nothing wrong with using a casual or conversational tone in technical writing, so long as it does not degenerate into slang. For example, this sentence is really going too far. Never use idioms like "crop up", "lose track", "it turned out that", "play up", or "right out".

Technical writing is not a good outlet for artistic impulses. The following is from a commercial software requirements document.

✗ The system should be developed with the end users clearly in view. It must therefore run the gamut from simplicity to sophistication, robustness to flexibility, all in the context of the individual user. From the first tentative familiarization steps, the consultation process has been used to refine the requirements by continued scrutiny and rigorous analysis until, by some alchemical process, those needs have been transmuted into specifications. These specifications distill the quintessence of the existing system.

The above extract has the excuse that it forms part of a sales pitch. But the following is from a scientific paper on concurrent database systems.

✗ We have already seen, in our consideration of *what is*, that the usual simplified assumptions lead inexorably to a representation that is desirable, because a solution is always desirable; but repugnant, because it is false. And we have presented *what should be*, assumptions whose nature is not susceptible to easy analysis but are the only tenable alternative to ignorance (absence of solution) or a false model (an incorrect solution). Our choice is then Hobson's choice, to make do with what material we have—viable assumptions—and to discover whether the intractable can be teased into a useful form.

Deciphering this paper was hard work. The following is a rough translation, but with no guarantee that the intended meaning is preserved.

✓ We have seen that the usual assumptions lead to a tractable model, but this model is only a poor representation of real behaviour. We therefore proposed better assumptions, which however are difficult to analyze. Now we consider whether there is any way in which our assumptions can be usefully applied.

Novice writers are sometimes tempted to imitate the style of, not science writing, but popular science writing.

✗ As each value is passed to the server, the "heart" of the system, it is checked to see whether it is in the appropriate range.

✓ Each value passed to the central server is checked to see whether it is in the appropriate range.

Don't dress up your ideas as if they were on sale. In the following I have changed the author's name to "Grimwade".

✗ Sometimes the local network stalls completely for a few seconds. This is what we call the "Grimwade effect", discovered serendipitously during an experiment to measure the impact of server configuration on network traffic.

✓ Sometimes the local network stalls for a few seconds. We first noticed this effect during an experimental measurement of the impact of server configuration on network traffic.

Examples

Use an example whenever it adds clarification. A small example often means the difference between communication and confusion, particularly if the concept being illustrated is fundamental to understanding of the paper. Each example should be an illustration of one concept; if you don't know what an example is illustrating, change it.

Examples can give substance to abstract concepts.

✓ In a semi-static model, each symbol has an associated probability representing its likelihood of occurrence. For example, if the symbols are characters in text then a common character such as "e" might have an associated probability of 12%.

Motivation

Many authors take considerable trouble over the structure of their articles but don't make it obvious to the reader. Not only should the parts of a paper be ordered in a logical way, but this logic should be clearly communicated.

The introduction will usually give some indication of the organization of the paper, by outlining the results and their basis, and may include a list of the parts of the paper, but these measures by themselves are not sufficient. Brief summaries at the start and end of each section are helpful, as are sentences linking one section to the next; for example, a well-written section might conclude with

✓ Together these results show that the hypothesis holds for linear coefficients. The difficulties presented by non-linear coefficients are considered in the next section.

Link text together as in a narrative—each section should have a clear story to tell. The connection between one paragraph and the next should be obvious.

A common error is to include material such as definitions or theorems without indicating why it is useful. Usually the problem is lack of explanation; sometimes it is symptomatic of an ordering problem, such as including material before the need for it is obvious. Never assume that a series of definitions, theorems, or algorithms—or even the need for the series—is self-explanatory. Motivate the reader at each major step in the exposition: explain how a definition (theorem, lemma, whatever) will be used, or why it is interesting, or how it fits into the overall plan.

The author of an article is almost always better informed than its readers. Even expert readers won't be familiar with some of the details of a problem, whereas the author has probably been studying the problem intimately for months or years and takes many difficult issues for granted. Authors should explain everything that is not common knowledge to the article's audience; what constitutes common knowledge depends on the article's subject and on where it is published.

When writing a paper you must decide what to teach the reader: a secret of good writing is to identify what the reader needs to learn. A paper is a sequence of concepts, building from a foundation of knowledge assumed to be common to all readers up to new ideas and results. At each part of a paper you should consider what the reader has learnt so far, whether this knowledge is sufficient to allow understanding of what follows, and whether each part is a natural consequence of what has already been said.

The upper hand

Some authors seem have a superiority complex—a need to prove that they know more or are smarter than their readers. Perhaps the most appropriate word for this behaviour is swagger. (Peter Medawar uses "scientmanship", to suggest the element of oneupmanship [23].) One form of swagger is implying familiarity with material that most scientists will never read; an example is reference to philosophers such as

Wittgenstein or Hegel, or statements such as "the argument proceeds on Voltarian principles". Another form is the unnecessary inclusion of difficult mathematics, or offhand remarks such as "analysis of this method is of course a straightforward application of tensor calculus". Yet another form is citation of obscure, inaccessible references.

This kind of showing off, of attempting to gain the upper hand over the reader, is snobbish and tiresome. Since the intention is to make statements the reader won't understand, the only information conveyed is an impression of the author's ego. Write for the dullest of your readers, as an equal.

Obfuscation

Obfuscation is the art of making statements in ambiguous or convoluted terms, with the intention of hiding meaning, or of appearing to say much while actually saying little. It is particularly useful when the aim is to give the impression of having done something without actually claiming to have done it.

✗ Experiments, with the improved version of algorithm as we have described, are the step that confirms our speculation that performance would improve. The previous version of the algorithm is rather slow on our test data and improvements lead to better performance.

Note the use of bland statements such as "experiments ... are the step that confirms our speculation" (true, but not exactly informative) and "improvements lead to better performance" (tautologous). The implication is that experiments were performed but there is no direct claim that experiments actually took place.

Deliberate obfuscation is not common in science writing, but vague writing is rife. It is always preferable to be specific: exceptions are or are not possible; data was transmitted at a certain rate; and so on. Stating that "there may be exceptions in some circumstances" or "data was transmitted fast" is simply not helpful.

Obfuscation can arise in other ways: exaggeration, omission of relevant information, bold statements of conclusions based on flimsy evidence. Use of stilted text—often an attempt to introduce unnecessary formality—can obfuscate.

✗ The status of the system is such that a number of components are now able to be operated.

✓ Several of the system's components are working.

✗ In respect to the relative costs, the features of memory mean that with regard to systems today disk has greater associated expense for the elapsed time requirements of tasks involving access to stored data.

✓ Memory can be accessed more quickly than disk.

Analogies

Analogies are curious things: what seems perfectly alike or parallel to one person may seem entirely unalike to another. Another drawback to analogies is that they can take your reasoning astray—two situations may have marked similarities but nonetheless have fundamental differences that the analogy leads you to ignore. For an analogy to be worthwhile, it should significantly reduce the work of understanding the concept being described. I have seen more bad analogies than good in computing research papers; however, simple analogies can undoubtably help illustrate unfamiliar concepts.

✓ Contrasting look-ahead graph traversal with standard approaches, look-ahead uses a bird's-eye view of the local neighbourhood to avoid dead ends—but at the cost of feeding the bird and waiting for it to return after each observation.

Straw men

A straw man is an indefensible hypothesis posed for the sole purpose of being demolished. A paraphrasing of an instance in a published paper is "it can be argued that databases do not require indexes", in which the author and reader are well aware that a database without an index is as practical as a library without a catalogue. Such writing says more about the author than it does about the subject.

Another form of straw man is the contrasting of a new idea with some impossibly bad alternative, to put the new idea in a favourable light. This form is obnoxious because it can lead the reader to believe that the impossibly bad idea might be worthwhile, and that the new idea is more important than is in fact the case. Contrasts should be between the new and the current, not the new and the fictitious.

✗ Query languages have changed over the years. For the first database systems there were no query languages and records were retrieved with programs. Before then data was kept in filing cabinets and indexes were printed on paper. Records were retrieved by getting them from the cabinets and queries were verbal, which led to many mistakes being made. Such mistakes are impossible with new query languages like QIL.

A more subtle form of straw man is comparison between the new and the ancient. For example, criticisms based on results in old papers are unreasonable because, in all likelihood, papers published since will have described improvements.

Reference and citation

Authors should carefully relate their new work to existing work, showing how their work builds on previous knowledge and how it differs from other, relevant results. The existing work is identified by reference to published articles, books, or reports. An article will include a bibliography, that is, a list of such references in a standardized format, and embedded in the article's text there will be citations to the publications.

References, and discussion of them, serve three main purposes. They help demonstrate that work is new: claims of originality are much more convincing in the context of references to existing work that (from the reader's perspective) appears to be similar. They demonstrate the author's knowledge of the research area, which is important if the author's statements are to be regarded as reliable. And they are pointers to background reading.

Before including a reference, consider whether it will be of service to the reader. A reference should be relevant, it should be up-to-date, it should be reasonably accessible, and it should be necessary—don't add citations just to pad the bibliography. Refer to an original paper in preference to a secondary source; well-written material in preference to bad; to a book or journal article in preference to a conference article; to a conference article in preference to a technical report or manuscript (which have the additional disadvantage of being unrefereed); and avoid reference to private communications and information provided in forums such as seminars or talks. Such information cannot be accessed or verified by the reader and references to it are not helpful. In the rare circumstance that you must refer to such material, do so via a footnote, parenthetical remark, or acknowledgement; don't put an entry in the bibliography.

Don't cite to support common knowledge. For example, use of a binary tree in an algorithm doesn't require a reference to a data structures text. But claims, statements of fact, and discussion of previous work should be substantiated by reference if not substantiated within the paper. This rule even applies to minor points—for some readers the minor points could be a major interest.

Some references will be the author's own papers. Such references establish the author's credentials as someone who understands the area, establish a research history for the article, and allow the interested reader to gain a deeper understanding of the research by following it from its inception. Gratuitous self-reference, however, undermines all of these purposes; it is frustrating for readers to discover that hard-sought references are not relevant. Technical reports in particular should not be self-referenced, unless they contain material that is genuinely important and not available elsewhere.

On rare occasions it is necessary to refer to a result in an inaccessible paper. For example, suppose that in 1981 Dawson wrote "Kelly (1959) shows that stable graphs are closed", but Kelly (1959) is inaccessible and Dawson (1981) does not give the details. In your article, do not refer directly to Kelly—after all, you can't check the details yourself, and Dawson may have made a mistake.

✓ According to Dawson (1981), stable graphs have been shown to be closed.

✓ According to Kelly (1959; as quoted by Dawson, 1981), stable graphs are closed.

The second form tells readers who originated the result without the effort of obtaining Dawson first. Kelly's entry in the bibliography should clearly show that the reference is second-hand.

Regardless of whether you have access to original sources, be careful to attribute results correctly. For example, there have been references to "Knuth's Soundex algorithm", although Knuth is not the author and the algorithm was at least fifty years old when Knuth discussed it.

Some readers of an article will not have access to the publications it cites, and so may rely on the article's description of these papers. For this reason alone you should describe results from other papers fairly and accurately. Any criticisms should be based on sound argument. Neither belittle papers, regardless of your personal opinion of their merits, nor overstate their significance; and beware of statements that might be interpreted as pejorative.

✗ Robinson's theory suggests that fast access is possible, but he did not perform experiments to confirm his results [22].

✓ Robinson's theory suggests that fast access is possible [22], but as yet there is no experimental confirmation.

Careful wording is needed in these circumstances. When referring to the work of Robinson, you might write that "Robinson thinks that ...", but this implies that you believe he is wrong, and has a faint odour of insult; you might write that "Robinson has shown that ...", but this implies that he is incontrovertably right; or you might write that "Robinson has argued that ...", but then should make clear whether you agree.

A simple method of avoiding such pitfalls is to quote from the reference, particularly if it contains a short, memorable statement—one or two sentences, say—that is directly pertinent. Quotation also allows you to clearly distinguish between what you are saying and what others have said, and is far preferable to plagiarism.

Cited material often uses a different terminology or notation, or is written for an entirely different context. When you use results from other papers, be sure to show the relationship to your own work. For example, a reference might show a general case, but you use a special case; then you need to show that is a special case. If you claim that concepts are equivalent, ensure that the equivalence is clear to the reader.

Unsubstantiated claims should be clearly noted as such, not dressed up as accepted facts.

✗ Most users prefer the graphical style of interface.

✓ We believe that most users prefer the graphical style of interface.

✗ Another possibility would be a disk-based method, but this approach is unlikely to be successful.

✓ Another possibility would be a disk-based method, but our experience suggests that this approach is unlikely to be successful.

Punctuation of citations is considered on page 67. Citation and quotation are discussed at length in Mary-Claire van Leunen's *Handbook for Scholars* [4].

Citation style

References that are discussed should not be anonymous.

✗ Other work [16] has used an approach in which ...

✓ Marsden [16] has used an approach in which ...
Other work (Marsden 1991) has used an approach in which ...

The latter versions provide more information to the reader, and "Marsden" is easier to remember than "[16]" if the same paper is discussed later on. Self-references particularly should not be anonymous—it should be clear to the reader that references used to support your argument are your own papers, not independent authorities. Other references that are not discussed can just be listed.

✓ Better performance might be possible with string hashing techniques that do not use multiplication [11, 30].

Avoid unnecessary discussion of references.

✗ Several authors have considered the problem of unbounded delay. We cite, for example, Hong and Lu (1991) and Wesley (1987).

✓ Several authors have considered the problem of unbounded delay (Hong and Lu 1991; Wesley 1987).

Two styles of citation are illustrated above. One is the ordinal-number style, in which entries in the reference list are numbered and are cited by their number, as in "... is discussed elsewhere [16]". The other is the name-and-date style—the "Harvard" style—in which entries are cited by author name, as in "... is discussed by Whelks and Babb (1972)" or "... is discussed elsewhere (Whelks and Babb 1972)". A third common style is to use superscripted ordinal numbers, as in "... is discussed elsewhere^{16}". My preference is for the ordinal-number styles because they encourage better writing.

Another style is use of uppercase abbreviations, where references are denoted by strings such as "[MAR91]". This is not a good style: the abbreviations tend to encourage poor writing such as "... is discussed in [WHB72]" and, because uppercase characters stand out from text, they are rather distracting.

Note however that many journals insist on a particular style. (Some also insist that entries be ordered alphabetically in the reference list, which is convenient for the reader; or that they appear by order of citation, which is convenient for typesetting.) Your writing should be designed to survive change in citation style.

When discussing a reference with more than three authors, all but the first author's name can be replaced by "et al."

✓ Howers, Mann, Thompson, and Wills [9] provide another example.

✓ Howers et al. [9] provide another example.

Note the stop: "et al." is an abbreviation.

Each entry in the reference list should include enough detail to allow readers to find the reference. Other than in extreme cases, the names of all authors should be given—it is preferable not to use "et al." in the reference list. An exception is the rare case in which the authors list themselves as "et al." (I have only seen one paper with such an author list: "The Story of O_2" by O. Deux et al.)

Format fields of the same type in the same way. For example, don't list one author as "Heinrich, J.", the next as "Peter Hurst", the next as "R. Johnson" and the next as "SL Klows". Capitalization, explained on page 64, should be consistent. Don't use unfamiliar abbreviations of journal names. (One that has puzzled me is "*J. Comp.*")

Journal articles. The journal name should be given in full and author names, paper title, year, volume, number, and pages must be provided. Consider also giving the month. Thus

✗ T. Wendell, "Completeness of open negation in quasi-inductive programs", *J. Dd. Lang.*, 34.

is inadequate; revise to, say,

✓ T. Wendell, "Completeness of open negation in quasi-inductive programs", *ICSS Journal of Deductive Languages*, 34(3):217–222, November 1994.

Conference articles. The conference name should be complete and authors, title, year, and pages must be provided. Information such as publisher, conference location, month, and editors should also be given.

Books. Give title, authors, publisher and publisher's address, year, and, where relevant, edition and volume. If the reference is to a specific part of the book, give page numbers; for example, write "(Howing 1994; pp. 22–31)" rather than just "(Howing 1994)". If the reference is to a chapter, give its title, pages, and, if applicable, authors.

Technical reports. In addition to title, authors, year, and report number you need to provide the address of the publisher (which is usually the authors' home institution). If the report is available online, say via ftp or http, consider giving its electronic address.

Obscure references. Take particular care to provide as much information as possible. If you must refer to the First Scandinavian Workshop on Backward Compatibility, consider explaining how to obtain the proceedings or a copy of the paper.

Quotation

Quotations are text from another source, usually included in a paper to support an argument. The copied text, if short, is enclosed in double quotes (which are more visible than single quotes and cannot be confused with apostrophes). Longer quotes are set aside in an indented block.

✓ Information retrieval is "the science of matching information needs to documents" (Brinton 1991).

✓ As described by Kanu [16], there are three stages:

> First, each distinct word is extracted from the data. During this phase, statistics are gathered about frequency of occurrence. Second, the set of words is analyzed, to decide which are to be discarded and what weights to allocate to those that remain. Third, the data is processed again to determine likely aliases for the remaining words.

The quoted material should be an exact transcription of the original text; some syntactic changes are permissible, so long as the meaning of the text is unaltered, but the changes should be held to a minimum. Changes of font, particularly addition of emphasis by changing words to italics, should be explicitly identified, as should changes of nomenclature.

The expression "[*sic*]" is used to indicate that an error is from the original quote, as in "Davis regards it as 'not worty [*sic*] of consideration' [11]". It is not polite to point out errors; avoid such use of "[*sic*]" and of quotes that seem to require it. More rarely "[*sic*]" is used to indicate that terminology or jargon is being used in a different way.

✗ Hamad (1990) shows that "similarity [*sic*] is functionally equivalent to identity"; note that similarity in this context means homology only, not the more general meaning used in this paper.

The long explanation renders the quote pointless.

✓ Hamad (1990) shows that homology "is functionally equivalent to identity".

For a short, natural statement of this kind the quotes are not really necessary.

✓ Hamad (1990) shows that homology is functionally equivalent to identity.

Other changes are insertions, replacements, or remarks, delimited by square brackets; and short omissions, represented by ellipses.

✓ They describe the methodology as "a hideous mess ... that somehow manages to work in the cases considered [but] shouldn't".

(Note that an ellipsis consists of three stops, neither more nor less.) Ellipses are unnecessary at the start of quotes, and at the end of quotes except where they imply "et cetera" or "and so on", or where the sentence is left hanging. For long omissions, don't use an ellipsis; separate the material into two quotations.

Don't mutilate quotations.

✗ According to Fier and Byke such an approach is "simple and ... fast, [but] fairly crude and ... could be improved" [8].

It would be better to paraphrase.

✓ Fier and Byke describe the approach as simple and fast, but fairly crude and open to improvement [8].

Long quotations, and quotation in full of material such as algorithms or figures, require permission from the publisher and from the author of the original.

Words can be quoted to show that they are inadequately defined.

✗ This language has more "power" than the functional form.

Here the author must assume that "power" will be understood in a consistent way by the reader. Such use of quotes indicates woolly thinking—that the author is not quite sure what "power" means, for example.

✓ This language allows simpler expression of queries than does the functional form.

More rarely, words can be quoted to indicate irony. The expression "in their 'methodology' " can be interpreted as *in their so-called methodology*, and is therefore insulting. I do not recommend these uses of quotes.

Acknowledgements

In the acknowledgements of a scientific paper you should thank everyone who made some contribution, be it advice, proofreading, or whatever: include research students, research assistants, technical support, and colleagues. Funding sources should also be acknowledged. It is usual to thank only those who contributed to the scientific content—don't thank your parents or your cat unless they really helped with the research. Books often have broader acknowledgements, to include thanks for people who have helped in non-technical ways. Consider showing your acknowledgement to the people you wish to thank, in case they object to the wording or to the presence of their name in the paper.

There are two common forms of acknowledgement. One is to simply list the people who have helped with the paper.

✓ I am grateful to Dale Washman, Kim Micale, and Dong Wen. I thank the Foundation for Science and Development for financial support.

Even in this little example there is some scope for bruised egos—Kim might wonder why Dale was listed first, for example.

The other common form is to explain each person's contribution. On the one hand, don't make your thanks too broad; if Kim and Dong constructed the proof, why aren't they authors? On the other hand, too much detail can damn with faint praise.

× I am grateful to Dale Washman for discussing aspects of the proof of Proposition 4.1, to Kim Micale for identifying technical errors in Theorem 3, and to Dong Wen for helping with some of the debugging. I thank the Foundation for Science and Development for financial support.

✓ I am grateful to Dale Washman and Kim Micale for our fruitful discussions, and to Dong Wen for programming assistance. I thank the Foundation for Science and Development for financial support.

This form has the advantage of identifying which of your colleagues contributed to the intellectual content.

Some authors write their thanks as "I would like to thank ..." or "I wish to thank ...". To me this seems to imply that *I wish to thank ... but for some reason I am unable to do so.* Consider instead using "I am grateful to ..." or simply "I thank ..."

Ethics

It can be tempting for authors to overstate the significance and originality of their results, and to diminish the status of previous results in the field, to increase the likelihood of their work being published. However, the scientific community expects that published research be new, objective, and fair. Authors should not present opinion as fact, distort truths, plagiarize others, or imply that previously published results are original.

Plagiarism includes, not just the direct copying of textual material, but the use of other people's ideas or results without acknowledgement. It can even be argued that authors who reuse their own text, once it has appeared in a copyright form such as a journal, are guilty of plagiarism. The issue has been considered by Stone[1] and by Samuelson.[2] Authors should usually write fresh text for each new paper.

[1] Harold S. Stone, "Copyrights and author responsibilities", *IEEE Computer*, 25(12):46–51, December 1992.

[2] Pamela Samuelson, "Self-plagiarism or fair use?", *Communications of the ACM*, 37(8):21–25, August 1994.

Publication of multiple papers based on the same results is widely regarded as improper unless there is full cross-referencing, for example by reference to a preliminary publication from a more complete article that is a later outcome of the same research. Simultaneous submission to more than one journal or conference of papers based on the same results should be disclosed at the time of submission; and the usual response to such a disclosure is to ask the author to withdraw the paper.

It is now common for authors to make their papers available via internet, most often as technical reports. Such informal publication opens two ethical issues. One is copyright: editors of journals or conferences might regard a paper that has been made available in this way as already published, and decline to consider it for formal publication. In the "ACM interim copyright policies",[3] such informal publication is not condemned, but once a paper has been accepted for formal publication it is expected that the informal version will be removed.

The other issue of online publication is permanency. When an author discovers an error in an online paper it is all to easy to correct it silently, with no explicit indication that the paper has changed. But readers can have no confidence that the paper hasn't changed in some essential way. (Such changes are of course also possible with a printed technical report, but the continued existence of the original version means that authors have a fixed document they can refer to.) Modifications to online papers should always be made explicit, by use of a version number and date of publication; and the original version should be kept available, as other authors may have referred to it.

Authorship

Deciding who has merited authorship of a paper can be a difficult and emotional issue. A broadly accepted view is that each author must have made some significant contribution to the intellectual content of the paper. Thus directed activities such as programming do not usually merit authorship, nor does proofreading. An author should have participated in the conception, execution, or interpretation of the results, and usually an author should have participated to some degree in all of these activities. But the point at which a contribution becomes "significant" is impossible to define, and every case is different.

[3] *Communications of the ACM*, 38(4):104–109, April 1995.

A researcher who has contributed to the research must be given an opportunity to be included as an author, but authors should not be listed without their permission.

A related issue is of author order, since many readers will assume that the first author is the main contributor. A researcher who is clearly the main contributor should always be listed first—don't believe Alfred Aaby when he tells you that alphabetic ordering is the norm. Where there is no obvious first author, possible approaches to ordering include alphabetical or reverse alphabetical, perhaps with an explanatory footnote, or a reversal or rotation of the order used on a previous paper by the same authors. Many supervisors choose to put their student co-authors first.

Grammar

In this book I have avoided giving advice on grammar, because the clarity of writing largely depends on whether it conforms to accepted usage. One aspect of grammar is, however, worth considering: that some people like to use traditional grammar to criticize other people's text, based on rules such as *don't split infinitives* or *don't begin a sentence with "and" or "but"*. I dislike this attitude to writing: grammatical rules should be observed, but not at the cost of clarity or meaning. However, be aware that an overdose of grammatical errors annoys some readers.

Beauty

Authors of style guides like to apply artistic judgements to text. This does not mean that scientific writing should be judged as literary prose, indeed such prose would be quite inappropriate. But we read that text should be crystalline, transparent, and have good rhythm and cadence; and one should dislike stuffiness, softness, stodge, sludge, and sagging or soggy sentences.

How useful such judgements are to most authors is not clear. Doubtless, well-crafted text is a pleasure to read, ill-written text can be hard going, and good rhythm in text helps us to parse. But awareness of beauty in text does not, I think, help us to attain it, nor is it evident that, to a poor writer, the terminology of beauty in text is meaningful. It is sufficient to aim to achieve simplicity and clarity.

3 Writing style: specifics

Those complicated sentences seemed to him very pearls ... "The reason for the unreason with which you treat my reason, so weakens my reason that with reason I complain of your beauty" ... These writings drove the poor knight out of his wits.

Cervantes
Don Quixote

Underneath the knocker there was a notice that said:

PLES RING IF AN RNSER IS REQIRD

Underneath the bell-pull there was a notice that said:

PLEZ CNOKE IF AN RNSR IS NOT REQID

These notices had been written by Christopher Robin, who was the only one in the forest who could spell.

A. A. Milne
Winnie the Pooh

Titles and headings

Titles of articles and sections should be concise and informative, use specific rather than general terms, and accurately describe the content. Complicated titles with long words are hard to swallow.

✗ A New Signature File Scheme based on Multiple-Block Descriptor Files for Indexing Very Large Data Bases

✓ Signature File Indexes Based on Multiple-Block Descriptor Files

✗ An Investigation of the Effectiveness of Extensions to Standard Ranking Techniques for Large Text Collections

✓ Extensions to Ranking Techniques for Large Text Collections

Don't make the title so short that it is contentless. "Limited-Memory Huffman Coding for Databases of Textual and Numeric Data" is awkward, but it is superior to "Huffman Coding for Databases", which is far too general.

Accuracy is more important than catchiness—"Strong Modes can Change the World!" is excessive, not to mention uninformative. The more interesting the title, however, the more likely that the text underneath it will be read. The title is the only part of your paper that the vast majority of people will see; if the title does not reflect the paper's contents, the paper will not be read by the right audience.

Titles and section headings do not have to be complete sentences; indeed, such titles can look rather odd.

✗ Duplication of Data Leads to Reduction in Network Traffic

✓ Duplicating Data to Reduce Network Traffic

Section headings should reflect the article's logical structure. If a section is headed "Lists and Trees" and the first subsection is "Lists", another should be "Trees"; don't use, say, "Other Data Structures". If a section is headed "Index Organizations" the subsection heading should be "B-trees" rather than "B-tree indexes".

An article (or thesis chapter) usually consists of sections and possibly subsections. There is rarely any need to break subsections into sub-subsections. Don't break text into small blocks; three headings on a page is almost certainly too many. But beware of having too few sections, because it is hard to continue the logical flow of a section over more than a few pages.

Headings may or may not be numbered. My preference is to use only two levels of headings, major and minor, and to only number major headings. If all headings are unnumbered, make sure that major and minor headings are clearly distinguished by font, size, or placement.

The opening paragraphs

The opening paragraphs can set the reader's attitude to the whole paper, so begin well. All of a document should be created and edited with care, but take the most care with the opening, to create the best possible impression. The abstract should be written especially well, without an unnecessary word, and the opening sentence should be direct and straightforward.

✗ Trees, especially binary trees, are often applied—indeed indiscriminately applied—to management of dictionaries.

✓ Dictionaries are often managed by a data structure such as a tree, but trees are not always the best choice for this application.

The following example of how not to begin is the first sentence of a published paper.

✗ This paper does not describe a general algorithm for transactions.

Only later does the reader discover than the paper describes an algorithm for a special case.

✓ General-purpose transaction algorithms guarantee freedom from deadlock but can be inefficient. In this paper we describe a new transaction algorithm that is particularly efficient for a special case, the class of linear queries.

The first paragraphs should be intelligible to any likely reader; save technicalities for later on, so that readers who can't understand the details of your paper are still able to understand your results and the importance of your work. That is, describe what you have done without the details of how it was done.

Starting an abstract or introduction with "This paper concerns ..." or "In this paper ..." often means that results are going to be stated out of context.

✗ In this paper we describe a new programming language with matrix manipulation operators.

✓ Most numerical computation is dedicated to manipulation of matrices, but matrix operations are difficult to implement efficiently in current high-level programming languages. In this paper we describe a new programming language with matrix manipulation operators.

The second version describes the context of the article's contribution.

A typical organization for the introduction of an article would be to use the first paragraphs to describe the context. It is these paragraphs that convince the reader that the article is likely to be interesting. The opening sentences should clearly indicate the topic.

✗ Underutilization of main memory impairs the performance of operating systems.

✓ Operating systems are traditionally designed to use the least possible amount of main memory, but such design impairs their performance.

The second version is better for several reasons. It is clear; it states the context, which can be paraphrased as *operating systems don't use much memory*; and, in contrast to the first version, it is positive.

Take care to distinguish description of existing knowledge from the description of the paper's contribution.

✗ Many user interfaces are confusing and poorly arranged. Interfaces are superior if developed according to rigorous principles.

✓ Many user interfaces are confusing and poorly arranged. We demonstrate that interfaces are superior if developed according to rigorous principles.

Don't write the introduction as if it flows on from the abstract, which is a summary of a paper rather than its opening. The paper should be complete even with the abstract removed.

Variation

Diversity—in organization, structure, length of sentences and paragraphs, and choice of words—is a useful device for keeping the reader's attention.

✗ The system of rational numbers is incomplete. This was discovered 2000 years ago by the Greeks. The problem arises with squares whose sides are of unit length. The length of the diagonals of these squares is irrational. This discovery was a serious blow to the Greek mathematicians.

✓ The Greeks discovered 2000 years ago that the system of rational numbers is incomplete. The problem is that some quantities, such as the length of the diagonal of a square with unit sides, are irrational. This discovery was a serious blow to the Greek mathematicians.

Note how, in the second version, the final statement is more effective although it hasn't been changed.

Paragraphing

A paragraph usually consists of discussion on a single topic or issue. In a well-written paper, the gist if not the argument is often captured in the first sentence of each paragraph, with the remainder of the paragraph used for amplification or example. Every sentence in a paragraph should be related to the topic announced in the opening.

Long paragraphs can be an indication that several lines of argument have not been sufficiently disentangled by the author. Moreover, readers tend to pay more attention to the start and end of each paragraph and less to the body. If a long paragraph can be broken, break it. Variation in paragraph length makes the page less dull in appearance, however, so don't chop text into paragraphs of uniform size.

Contextual information can be forgotten between paragraphs, and references between paragraphs can be difficult to follow. For example, if a paragraph discusses a fast sorting algorithm, the next paragraph should not begin "This algorithm ..." but rather "The fast sorting algorithm ..."; if one paragraph refers to Harvey, the next should not refer to "his" but rather "Harvey's". Link paragraphs by reuse of key words or phrases, and by using expressions that connect the content of one paragraph to that of the next.

The use of formatted lists as an occasional alternative to paragraphs is common. Lists are useful for the following reasons.

- They highlight each main point clearly.

- The context remains obvious, whereas in a long list of points made in a paragraph it is hard to tell whether the later points are part of the original issue or belong to some subsequent discussion.
- An individual point can be considered in detail without confusing the main thread of narrative.
- They are easy to refer to; for example, as a checklist of the necessary properties of an algorithm.

List points can be numbered, named, or tagged. Use numbers only when ordering is important. If it is necessary to refer to an individual point, use numbers or names. Otherwise use tags, as in the list above. Acceptable tags are bullets and dashes; fancy symbols such as $\hookrightarrow$ or graphic icons look childish.

A disadvantage of lists is that they highlight rather too well: a list of trivia can be more attention-getting than a paragraph of important information. Reserve the use of lists for material that is both important and in need of enumeration.

Sentence structure

Sentences should have simple structure, which usually means that they will be no more than a line or two. Don't say too much all at once.[4]

[4]The following quote is a single sentence from a version of the standard lease agreement of the Real Estate Institute of Victoria, Australia. It is 477 words long, but the punctuation amounts to only three pairs of parentheses, one comma, and one stop. This clause is an example of "the fine print"—for example, the holder of a lease containing this clause has agreed not to take action if, in circumstances such as failure to pay rent, assaulted by the property's owner.

> If the Lessee shall commit a breach or fails to observe or perform any of the covenants contained or implied in the Lease and on his part to be observed and performed or fails to pay the rent reserved as provided herein (whether expressly demanded or not) or if the Lessee or other person or persons in whom for the time being the term hereby created shall be vested, shall be found guilty of any indictable offence or felony or shall commit any act of bankruptcy or become bankrupt or make any assignment for the benefit of his her or their creditors or enter into an agreement or make any arrangement with his her or their creditors for liquidation of his her or their debts by composition or otherwise or being a company if proceedings shall be taken to wind up the same either voluntarily or compulsorily under any Act or Acts relating to Companies (except for the purposes of reconstruction or amalgamation) then

✗ When the kernel process takes over, that is when in the default state, the time that is required for the kernel to deliver a message from a sending application process to another application process and to recompute the importance levels of these two application processes to determine which one has the higher priority is assumed to be randomly distributed with a constant service rate R.

✓ When the kernel process takes over, one of its activities is to deliver a message from a sending application process to another application process, and to then recompute the importance levels of these two application processes to determine which has the higher priority. The time required for this activity is assumed to be randomly distributed with a constant service rate R.

That the kernel process is the default state is irrelevant here, and should have been explained elsewhere.

This example also illustrates the consequence of having too many words between related phrases. The original version said that "the time that is required for *something* is assumed to be ...", where *something*

---

and in any of the said cases the Lessor notwithstanding the waiver by the Lessor of any previous breach or default by the Lessee or the failure of the Lessor to have taken advantage of any previous breach or default at any time thereafter (in addition to its other power) may forthwith re-enter either by himself or by his agent upon the Premises or any part thereof in the name of the whole and the same have again repossess and enjoy as in their first and former estate and for that purpose may break open any inner or outer doorfastening or other obstruction to the Premises and forcibly eject and put out the Lessee or as permitted assigns any transferees and any other persons therefrom and any furniture property and other things found therein respectively without being liable for trespass assault or any other proceedings whatsoever for so doing but with liberty to plead the leave and licence which is hereby granted in bar of any such action or proceedings if any such be brought or otherwise and upon such re-entry this Lease and the said term shall absolutely determine but without prejudice to the right of action of the Lessor in respect of any antecedent breach of any of the Lessee's covenants herein contained provided that such right of re-entry for any breach of any covenant term agreement stipulation or condition herein contained or implied to which Section 146 of the Property Law Act 1958 extends shall not be exercisable unless and until the expiration of fourteen days after the Lessor has served on the Lessee the Notice required by Sub-section(1) of the said Section 146 specifying the particular breach complained of and if the breach is capable of remedy requiring the Lessee to remedy the breach and make reasonable compensation in money to the satisfaction of the Lessor for the breach.

was 34 words long. The main reason that the revision is clearer is that *something* has been reduced to two words; the structure of the sentence is much easier to see.

It is likewise helpful to avoid nested sentences, that is, information embedded within a sentence that is not part of its main statement.

✗ In the first stage, the backtracking tokenizer with a two-element retry buffer, errors, including illegal adjacencies as well as unrecognized tokens, are stored on an error stack for collation into a complete report.

First, this is poor because crucial words are missing; the beginning should read "In the first stage, which is the backtracking tokenizer ...". Second, the main information—how errors are handled—is intermixed with definitions. Nested content, particularly if in parentheses, should be omitted. If it really is required then put it in a separate sentence.

✓ The first stage is the backtracking tokenizer with a two-element retry buffer. In this stage possible errors include illegal adjacencies as well as unrecognized tokens; when detected, errors are stored on a stack for collation into a complete report.

Watch out for fractured "if" expressions.

✗ If the machine is lightly loaded then speed is acceptable whenever the data is on local disks.

✓ If the machine is lightly loaded and data is on local disks then speed is acceptable.

✓ Speed is acceptable when the machine is lightly loaded and data is on local disks.

The first version is poor because the conditions of the "if" have been separated by the consequent.

It is easy to construct long, winding sentences by, for example, stating a principle, then qualifying it—a habit that is not necessarily bad, but does often lead to poor sentence structure—then explaining the qualification, the circumstances in which it applies, and in effect allowing the sentence to continue to another topic, such as the ideas underlying the principle, cases in which the qualification does or does not apply, or material which no longer belongs in the sentence at all; a property that is arguably true of most of this sentence, which should definitely be revised.

Sometimes longer sentences can be divided by, say, simply replacing an "and" or a semi-colon with a period. If there is no particular reason to join two sentences, keep them separate.

Beware of misplaced modifiers.

✗ We collated the responses from the users, which were usually short, into the following table.

✓ The users' responses, most of which were short, were collated into the following table.

Double negatives are difficult to parse and are often ambiguous.

✗ There do not seem to be any reasons not to adopt the new approach.

The impression here is of condemnation—*we don't like the new approach but we're not sure why*—but praise was intended; the quote is from a paper advocating the new approach. This is another example of the academic tendency to overqualification. The revision "There is no reason not to adopt the new approach" is punchier, but still negative. It is difficult to suggest further improvement with the same meaning, because the meaning was probably unintended; the following better reflects the authors' aims.

✓ The new approach is at least as good as the old and should be adopted.

Sing-song phrases are distracting, as are rhymes and alliteration.

✗ We propose that the principal procedure of proof be use of primary predicates.

✗ Semantics and phonetics are combined by heuristics to give a mix that is new for computational linguistics.

Repetition and parallelism

Text that consists of the same form of sentence used again and again is monotonous. Watch out for sequences of sentences beginning with "however", "moreover", "therefore", "hence", "thus", "and", "but", "then", "so", "nevertheless", or "nonetheless". Likewise, don't overuse the pattern "First, ... Second, ... Last, ..."

Complementary concepts should be explained as parallels, or the reader will have difficulty seeing how the concepts relate.

✗ In SIMD, the same instructions are applied simultaneously to multiple data sets, whereas in MIMD different data sets are processed with different instructions.

✓ In SIMD, multiple data sets are processed simultaneously by the same instructions, whereas in MIMD multiple data sets are processed simultaneously by different instructions.

Parallels can be based on antonyms.

✗ Access is fast, but at the expense of slow update.

✓ Access is fast but update is slow.

Lack of parallel structure can result in ambiguity.

✗ The performance gains are the result of tuning the low-level code used for data access and improved interface design.

✓ The performance gains are the result of tuning the low-level code used for data access and of improved interface design.

This can be improved again: it is kinder to the reader to move the longer clauses in a list to the end.

✓ The performance gains are the result of improved interface design and of tuning the low-level code used for data access.

There are some standard forms of parallel. The phrase "on the one hand" should have a matching "on the other hand". A sentence beginning "One ..." suggests that a sentence beginning "Another ..." is imminent. If you flag a point with "First" then every following point should have a similar flag, such as "Second", "Next", or "Last".

Parallel structures should be used in lists.

✗ To achieve good performance there should be sufficient memory, parallel disk arrays should be used, and caching.

The syntax can be fixed by adding "should be used" at the end but the result is clumsy. A complete revision is preferable.

✓ Achievement of good performance requires sufficient memory, parallel disk arrays, and caching.

Direct statements

Avoid excessive use of indirect statements (also known as passive voice), particularly descriptions of actions that don't include any indication of who or what performs the actions.

✗ The following theorem can now be proved.

✓ We can now prove the following theorem.

The direct style (or active voice) is often less stilted and easier to read.

Another unpleasant indirect style is the artificial use of verbs like "perform" or "utilize", in the false belief that such writing is more precise or scientific.

✗ Tree structures can be utilized for dynamic storage of terms.

✓ Terms can be stored in dynamic tree structures.

✗ Local packet transmission was performed to test error rates.

✓ Error rates were tested by local packet transmission.

Other words often used in this way include "achieved", "carried out", "conducted", "done", "occurred", and "effected".

Change of voice sometimes changes meaning and often changes emphasis. If passive voice is necessary, use it. Complete absence of active voice is unpleasant, but that does not mean that all use of passive voice is poor.

Use of "we" is valuable when trying to distinguish between the contribution made in an article and existing results in a field, especially in an abstract or introduction. For example, in "it is shown that stable graphs are closed" the reader may have difficulty deciding who is doing the showing, and in "it was hypothesized that ..." the reader will be unsure whether the hypothesis was posed in this article or another one. Use of "we" can also allow some kinds of statements to be made more simply—consider "we show" versus "in this paper it is shown that". And "we" is preferable to pretentious expressions such as "the authors".

Some authors use phrases such as "this paper shows ..." and "this section argues ..." These phrases, with their implication that the paper is sentient, should not be used.

In some cases the use of "we" is wrong.

✗ When we conducted the experiment it showed that our conjecture was correct.

Here, the use of "we" suggests that if someone else ran the experiment it would behave differently.

✓ The experiment showed that our conjecture was correct.

I do not particularly like the use of "I" in science writing, except when it is used to indicate that what follows is the author's opinion. The use of "I" in place of "we" in papers with only one author is, however, becoming more common.[5]

Ambiguity

Check carefully for ambiguity. It is often hard to detect in your own text because you know what is intended.[6]

✗ The compiler did not accept the program because it contained errors.

✓ The program did not compile because it contained errors.

The next example is from a manual.

---

[5]Use of personal pronouns is a contentious issue in science writing. Some feel that it undermines objectivity by introducing the author's personality and is therefore unacceptable, even unscientific. Others argue that to suggest that a paper is not the work of individuals is intellectually dishonest, and that in any case use of personal pronouns makes papers easier to read.

Kirkman [16] has surveyed the use of personal statements in technical writing. Respondents were asked to comment on stylistically varied versions of the same text. From some 2 800 responses, clear preference was shown for writing that makes use of personal constructions, with more formal versions described as laborious and difficult, among other things. Interestingly, many respondents described the more personal, chatty style as unambiguous and easier to understand than the alternatives—yet felt it to be not proper for scientific writing.

[6]A safe-sex guide issued by the Australian Government included "a table on which sexual practices are safe"; it transpired that this was not a piece of furniture. Newspaper headlines can be a rich source of ambiguity:

> Enraged Cow Injures Farmer with Axe
> Miners Refuse to Work after Death

While not exactly ambiguous, the report that

> the pilot of a plane that crashed killing six people was flying "out of his depth"

does convey rather the wrong impression.

✗ There is a new version of the operating system, so when using the "fetch" utility, the error messages can be ignored.

✓ There is a new version of the operating system, so the "fetch" utility's error messages can be ignored.

Part of the confusion comes from the redundant phrase "when using": there would be no error messages if the utility was not being used.

Ensure that pronouns such as "it", "this", and "they" have a clear referent.

✗ In addition to skiplists we have also tried trees. They are superior because they are slow in some circumstances but have lower asymptotic cost.

✓ In addition to skiplists we have also tried trees. Skiplists are superior because, although slow in some circumstances, they have lower asymptotic cost.

A common source of confusion is between speed and time. Although not ambiguous, the phrase "increasing speed" is easily read as *increasing time*, which has quite the opposite meaning. There are similar problems with phrases such as "improving affordability".

A clumsy sentence is usually preferable to an ambiguous one. But remember that stilted sentences bog the reader down, and it is almost impossible to entirely avoid ambiguity.[7]

Emphasis

The structure of a sentence places implicit emphasis, or stress, on some words; reorganizing a sentence will change the emphasis.

---

[7]The following my-dog-has-no-nose joke, due to Andy Clews, is not ambiguous.

"First circumlocutionist: I have in my possession a male animal belonging to the family Canidae, and it appears that he does not possess any extra-facial olfactory organs.

Second circumlocutionist: Could you therefore impart to me such knowledge as is necessary to describe how that animal circumvents the problem of satisfying his olfactory senses?

First circumlocutionist: Unfortunately the non-ambiguity of your enquiry does not easily permit me to provide a clever answer, but I am in fact thinking of referring the animal to an olfactologist. However, the animal does have an unpleasant body odour, should you be interested."

✗ The algorithm is appropriate because each item is written once and read often.

It is not clear what makes the algorithm's behaviour appropriate; the emphasis should be on the last two words, not the last five.

✓ The algorithm is appropriate, because each item is only written once but is read often.

Inappropriate stress can lead to ambiguity.

✗ Additional memory can lead to faster response, but user surveys have indicated that it is not required.

✓ Faster response is possible with additional memory, but user surveys have indicated that it is not required.

The first version, which has the stress on "additional memory", incorrectly implies that users had commented on memory rather than response; and since the sentence is about "response", that is where the stress should be.

Explicit stress can be provided with italics, but is almost never necessary. Don't italicize words *unnecessarily*—let sentence structure provide the emphasis. Few papers require explicit stress more than once or twice. DON'T use capitals for emphasis. Some authors use the word "emphatic" to provide emphasis, as in "which are emphatically not equivalent"; another word used in this way is "certainly". The resulting wordiness weakens rather than strengthens; it is certainly not a good idea.

Italicized passages of any length are hard to read. Rather than italicize a whole sentence, say, stress it in some other way: italicize one or two words only, or make it the opening sentence of a paragraph.

The first time a key word is used, consider placing it in italics.

✓ The data structure has two components, a *vocabulary* containing all of the distinct words and, for each word, a *hit list* of references.

Definitions

Terminology, variables, abbreviations, and acronyms should be defined or explained when they first appear. Use a consistent format for introducing new terminology; implicit or explicit emphasis on the first occurrence of a new word is often helpful, because it stresses what is being introduced.

✗ We use homogeneous sets to represent these events.

This is poor because the reader does not know that "homogeneous" is a new term that is about to be defined, and may look back for an explanation.

✓ We use *homogeneous* sets to represent these events.

✓ To represent these events we use homogeneous sets, whose members are all of the same type.

It can be helpful to explain concepts twice, in different ways.

✓ Compaction, in contrast to compression, does not preserve information; that is, compacted data cannot be exactly restored to the original form.

Sometimes a discursion is needed to motivate a definition. The discursion might be of negative examples, showing what happens in the absence of the definition, or it might lead the reader by steps to agree that the entity being defined is necessary.

Choice of words

There is a strong trend in current writing towards using short, direct words rather than long, circumlocutionary ones; the result is vigorous, emphatic writing. For example, use "begin" rather than "initiate", "first" rather than "firstly", "part" rather than "component", and "use" rather than "utilize". Use short words in preference to long, but use an exact long word rather an approximate short one.

Use words that are specific and familiar. Abstract, vague, or broad terms will have different meanings to different readers and are more likely to lead to confusion.

✗ The analysis derives information about programs.

The "information" could be anything: optimizations, function-point analysis, bug reports, complexity.

✓ The analysis estimates the resource costs of programs.

Other abstract terms that are overused are "method" and "performance". "Difficult" is often used when a better term is available: if something is "difficult to compute", does that mean that it is slow, or memory-hungry, or requires double precision, or something else altogether? "Efficient" is another word that is often vague. Always use the most precise term available.

A common reason for using vague terms is that some authors feel they are writing badly if they use the same word twice in a sentence or paragraph, and thus substitute a synonym, which is usually less specific.

✗ The database executes on a remote machine to provide better security for the system and insulation from network difficulties.

✓ The database executes on a remote machine to provide better security for the database and insulation from network difficulties.

The "don't repeat words" rule might apply to descriptive writing, but not to technical terms that must be clearly understood.

Language is not static; words enter the language, or go out of vogue, or change in meaning (although this does not mean that we should deliberately debase or misuse words). A word whose meaning has changed—at least, some people still use the old meaning, but most use the new—is "data". Since "data" is by etymology a plural, expressions such as "the data is stored on disk" are grammatically incorrect, but "the data are stored on disk" simply seems wrong. Correspondingly, "datum" is now rare. "Data" is appropriate for both singular and plural. On the other hand, use "automaton" rather than "automata" for the singular case.

Use a word only if you are sure that you know the meaning and can apply it correctly. Some words are familiar because of their use in a certain context—perhaps in a saying such as "hoist by his own petard"—but have otherwise lost their meaning. Other words, such as "notwithstanding", have an archaic feel and seem out of place in new writing. Some words have acquired meanings in computing that are distinct from their meaning in general usage. Besides reuse of nouns such as "bus" there are more subtle cases. For example, "iterate" in computing means *to loop*, but in other writing it can mean *to do again*.

Slang words should never be used in technical writing. Nor should the choice of words suggest that the writing is careless; avoid sloppy-looking abbreviations and contractions. Use "cannot" in preference to "can't", for example.

Don't make excessive claims about your own ideas and results, or describe them with superlatives. Phrases such as "our method is an ideal solution to ..." or "our results are startling" are unacceptable. Claims about your own work should be unarguable.

Qualifiers

Qualifiers shouldn't be piled on top of each other. Words such as "might", "may", "perhaps", "possibly", "likely", "likelihood", and "could" can be used once in a sentence, but not more. Overuse of qualifiers results in text that is lame and timid.

✗ It is perhaps possible that the algorithm might fail on unusual input.

✓ The algorithm might fail on unusual input.

✓ It is possible that the algorithm would fail on unusual input.

Here is another example, from the conclusions of a paper.

✗ We are planning to consider possible options for extending our results.

✓ We are considering how to extend our results.

Double negatives are a form of qualifier; they are commonly used to express uncertainty.

✗ Merten's algorithm is not dissimilar to ours.

Such statements tell the reader little.

Qualifiers such as "very" and "quite" should be avoided altogether, because they are in effect meaningless. If an algorithm is "very fast", is an algorithm that is merely "fast" deficient in some way? Text is invariably more forceful without "very".

✗ There is very little advantage to the networked approach.

✓ There is little advantage to the networked approach.

Likewise, "simply" can often be deleted.

✗ The standard method is simply too slow.

✓ The standard method is too slow.

Padding

Padding is the use of pedantic phrases such as "the fact that" or "in general" which should be deleted, not least because they are irritating. The phrase "of course" can sound patronizing or even insulting—"*of course* the reader has observed that ..." The phrase "note that" is not padding, but should only be used to introduce something that readers can deduce for themselves, such as a consequence of a definition.

Phrases involving the word "case" ("in any case", "it is perhaps the case") are also suspect. There is no reason to use "it is frequently the case that ..." instead of "often ...", for example.

Unnecessary introduction of quantities, or the concept of quantities, is a form of padding. For example, the phrase "a number of" can be replaced by "several", and "a large number of" by "many".

Misused words

The first table on page 49 lists some words that are often spelt incorrectly in science writing. The second table lists words that are often used incorrectly because of confusion with another word of similar form or sound. The "usually correct" form is shown on the left; the form with which each word gets confused is shown on the right.

A problem word with regard to spelling is "disk"; both this spelling and "disc" are so common that neither is preferred, but be consistent. (Errors introduced by global substitution of one for the other are not always easy to diskover.) Other words that don't have a stable spelling include "enquire" ("inquire"), "biased" ("biassed"), and "dispatch" ("despatch"). Whether you use "-ise" or "-ize" in words such as "optimize" depends on the country in which you are writing. Given the international nature of science, which you use is unimportant but be consistent. Note that "ae" is obsolete in many words: "encyclopaedia" has become "encyclopedia", for example.

Some other problem words are as follows.

Which, that, the. Many writers use "which" when they mean "that". Use "that" in preference to "which"; use "which" when it cannot be replaced by "that".

✗ There is one method which is acceptable.

Right	*Wrong*	*Right*	*Wrong*
adaptation	adaption	miniature	minature
apparent	apparant	occasional	occaisional
argument	arguement	occurred	occured
consistent	consistant	participate	particepate
definite	definate	preceding	preceeding
existence	existance	primitive	primative
foreign	foriegn	propagate	propogate
grammar	grammer	referred	refered
heterogeneous	heterogenous	separate	seperate
homogeneous	homogenous	supersede	supercede
independent	independant	transparent	transparant
insoluble	insolvable		

Misspelt words

Usual	*Other*	*Usual*	*Other*
alternative	alternate	foregoing	forgoing
comparable	comparative	further	farther
complement	compliment	elusive	illusive
dependent	dependant	manyfold	manifold
descendant	descendent	omit	emit
discrete	discreet	partly	partially
emit	omit	principle	principal
ensure	insure	simple	simplistic
excerpt	exert	solvable	soluble

Misused words

✓ There is one method that is acceptable.

✓ There are three options, of which only one is tractable.

The word "that" is often underused. Use of "that" can make a sentence seem stilted, but its absence can make the sentence unclear.

✗ It is true the result is hard to generalize.

✓ It is true that the result is hard to generalize.

On the other hand, "the" is often used unnecessarily; delete it where doing so does not change the meaning.

May, might, can. Many writers use "may" or "might" when they mean "can". Use "may" to indicate personal choice, and "can" to indicate capability.

✓ Users can access this facility, but may not wish to do so.

Less, fewer. Use "less" for continuous quantities ("it used less space") and "fewer" for discrete quantities ("there were fewer errors").

Affect, effect. The "effect", or *consequence*, of an action is to "affect", or *influence*, outcomes.

Alternate, alternative, choice. The word "alternate" means *other* or *switch between*, whereas an "alternative" is something that can be chosen. If there is but one alternative, there is no choice; "alternative" and "choice" are not synonyms.

Basic, fundamental, sophisticated. Some writers confuse "basic" with "fundamental": the former means *elementary* as well as *a foundation*. A result should only be described as "basic" if *elementary* is meant, or readers may get the wrong idea. Likewise "sophisticated" does not mean *new*.

Conflate, merge. The word "conflate" means *regard distinct things as similar*; while "merge" means *join distinct things to form one new thing*. They are not equivalent.

Continual, continuous. "Continual" is not equivalent to "continuous". The former means *ceaselessly*, the latter means *unbroken*.

Conversely, similarly, likewise. Only use "conversely" if the statement that follows really is the opposite of the preceding material. Don't use "similarly" or "likewise" unless whatever follows has a strong parallel to the preceding material.

Fast, quickly, presently, timely, currently. A process is "fast" if it *runs quickly*; "quickly" means *fast*, but does not necessarily mean *in the near future*. Something is "timely" if it is *opportune*; timeliness has nothing to do with rapidity. Also on the subject of time, "presently" means *soon*, whereas "currently" means *at present*.

Optimism, minimize, maximize. Absolute terms are often misused. One such word is "optimize", which means *find an optimum* or *find the best solution* but is often used to mean just *improve*. The latter usage is now so common that it could be argued that the meaning of "optimize" has changed, but as there is no synonym for "optimize" such a change would be unfortunate. Other absolute terms that are misused are "maximize" and "minimize".

Spelling conventions

The English-speaking countries have different spelling conventions. The most important discrepancy in spelling is between Britain and the United States. For example, the American "traveler" becomes the British "traveller" while "fulfill" becomes "fulfil". In Britain it is incorrect to spell "-our" words as "-or", but, for example, "vigour" and "vigorous" are both correct. The American "center" is the British "centre", "program" is "programme" (except for *computer program*), and "catalog" is "catalogue". Perhaps the greatest confusion is with regard to the suffixes "-ize" and "-yze", which have the same recommended spelling in both countries, but are often spelt as "-ise" and "-yse" outside the United States. (However, these problems are overrated. Of the 6000 or so distinct words used to write this book, for example, other than "-ize" words only 20 or so have a nationality-specific spelling.) British spelling has been used throughout this book.

Science is international—technical writing is usually for an audience that is accustomed to reading text from around the world—and it is accepted that a national of one country won't use the spelling of another. The most important thing is to spell consistently and to be consistent with suffixes such as "-ize". But note that many journals insist on their own standards for spelling and presentation, or insist that the spelling be consistently of one nationality or another, and thus may choose to modify anything they publish.

The best authority for national spelling is usually a respectable dictionary written for that country. Remember however that dictionaries are a record of current non-technical spelling, not prescribed spelling—the presence of a spelling in a dictionary does not prove that it is appropriate to a particular discipline. The choice of a particular spelling for a technical term may be dictated by the usual spelling in other articles, not the nationality of the writer.

Jargon

The word "jargon" means *terms used in a specialized vocabulary* or *mode of speech familiar only to a group or profession.*[8] As such, the use of jargon is an important part of scientific communication—how convenient it is to be able to say "CPU" rather than "the part of the computer that executes instructions". Some use of technical language, which will inevitably make the writing inaccessible to a wider audience, is essential for communication with specialists. But the more technical the language in an article, the smaller the audience will be.

In mathematical writing, formal notation is a commonly-used jargon. Mathematics is often unavoidable but that doesn't mean that it must be impenetrable.

✗ **Theorem.** Let $\delta_1, \ldots, \delta_n$, $n > 2$ be such that $\delta_1 \mapsto_{\Omega_1} \delta_2, \ldots, \delta_{n-1} \mapsto_{\Omega_{n-1}} \delta_n$. Let $\eta', \eta'' \in \mathcal{R}$ be such that $\Omega_1 \models \eta'$ and $\Omega_{n-1} \models \eta''$. Then

$$\exists(\eta', \eta_1)(\eta_1, \eta_2) \cdots (\eta_{r-1}, \eta_r)(\eta_r, \eta'') \in L$$

such that $\forall \eta_i$, $1 \leq i \leq r$, $\exists \Omega_j$, $1 \leq j < n$, such that $\Omega_j \models \eta_i$.

[8] To which meanings the Oxford Dictionary adds *unintelligible or meaningless talk or writing; nonsense; gibberish; twittering.*

Mathematics as jargon is discussed further on page 72.

Jargon does not have to consist of obscure terms, indeed it can be at its most confusing when words in common use are given a new meaning; and some words have multiple meanings even within computing.

✗ The transaction log is a record of changes to the database.

✓ The transaction log is a history of changes to the database.

The first version is potentially confusing because databases consist of records. Likewise, consider "the program's function ..." Such problems can also occur with synonyms.

✗ Hughes describes an array of algorithms for list processing.

✓ Hughes describes several algorithms for list processing.

New jargon inevitably arises in the research process, as ideas are debated and simple labels attached to newly familiar concepts. Authors need to consider whether the terminology they use has the intended meaning (or any meaning at all) for readers.

The need to name variants of existing ideas or systems presents a dilemma, because if the new name is dissimilar to the old then the relationship is not obvious, but prefixing a modifier to the old name—for example, to obtain "binary tree" from "tree"—can result in ridiculous constructs such as the "variable-length bitstring multiple-descriptor floating bucket extensible hashing scheme".

Where new terminology or jargon is introduced, be careful to use it consistently. Existing terminology or notation should only be changed with good reason. Sometimes your problem requires new terminology that is inconsistent with the terminology already being used, thus making change essential; but remember that any change is likely to make your paper harder for others to read.

Foreign words

If you use a foreign word that you feel needs to go in italics, consider instead using an English equivalent. Some writers feel that use of foreign words is *de rigueur* because it lends the work a certain *je ne sais quoi* and shows *savoir-vivre*, but such writing is hard to understand.

It is polite to use appropriate characters for foreign names. Don't write "Børstëdt" as "Borstedt", for example.

Overuse of words

Repetition of a word is annoying when it makes the reader feel they have read the same phrase twice, or have read a phrase and an inversion of it.

✗ Ada was used for this project because the underlying operating system is implemented in Ada.

✓ Ada was used for this project because it is the language used for implementation of the underlying operating system.

Repetition should be eliminated when the same word is used in different senses, or when a word and a synonym of it are used together.

✗ Values are stored in a set of accumulators, each initially set to zero.

✓ Values are stored in a set of accumulators, each initialized to zero.

Some words are memorable and grate when they are used too frequently. Common offenders include "this", "very", and "also". Other words are even more memorable—unusual words, other than necessary technical terms, should only be used once or twice in a paper. Watch out for tics: heavy use of some stock word or phrase. Common tics include "so", "also", "hence", "note that", and "thus".

There are cases in which repetition is useful. In the phrase "discrete quantities and continuous quantities", the first "quantities" can be omitted, but such omissions are ambiguous surprisingly often and can result in text that is difficult to parse. What, for example, is intended by "from two to four hundred"? A common error relating to this form of expression is to shorten phrases by deleting adjectives, such as the second "long" in the expression "long lists and long arrays". Technical concepts should always be described in the same way, not by a series of synonyms.

Redundancy and wordiness

Use the minimum number of words, of minimum length, in your writing. The table on page 55 lists common redundant or wordy expressions and possible substitutes for them. The list is illustrative rather than exhaustive; there are some typical forms of redundancy, such as "completely unique" for "unique", for which there are hundreds of examples.

Bad	*Good*
adding together	adding
after the end of	after
in the region of	approximately
cancel out	cancel
conflated together	conflated
cooperate together	cooperate
currently ... today	currently ...
divided up	divided
give a description of	describe
during the course of	during
totally eliminated	eliminated
of fast speed	fast
first of all	first
for the purpose of	for
in view of the fact	given
joined up	joined
of large size	large
semantic meaning	meaning
merged together	merged
the vast majority of	most
it is frequently the case that	often
completely optimized	optimized
separate into partitions	partition
at a fast rate	quickly
completely random	random
reason why	reason
a number of	several
cost in size	size
such as ... etc.	such as ...
completely unique	unique
in the majority of cases	usually
whether or not	whether
the fact that	—
it can be seen that	—
it is a fact that	—

Examples of redundant or wordy expressions

Tense

In science writing, most text is in past or present tense. Present tense is used for eternal truths. Thus we write "the algorithm has complexity $O(n)$", not "the algorithm had complexity $O(n)$". Present tense is also used for statements about the text itself. It is better to write "related issues are discussed below" than to write "related issues will be discussed below".

Past tense is used for describing work and outcomes. Thus we write "the ideas were tested by experiment", not "the ideas are tested by experiment". It follows that occasionally it is correct to use past and present tense together.

✓ Although the algorithm has worst-case complexity $O(n^2)$, in our experiments the worst case observed was $O(n \log n)$.

Either past or present tense can be used for discussion of references. Present tense is preferable but past tense can be forced by context.

✓ Willert shows that the space is open [14].

✓ Haast postulated that the space is bounded [7], but Willert has since shown that it is open [14].

Other than in conclusions, future tense is rarely used in science writing.

Plurals

When describing classes of things, excessive use of plurals can be confusing. Consider the following, adapted from a paper on minimum redundancy codes.

✗ Packets that contain an error are automatically corrected.

✗ Packets that contain errors are automatically corrected.

The first version implies that packets with a particular error are corrected, the second that packets with multiple errors are corrected. Both of these interpretations are wrong. Whenever it is reasonable to do so, convert plurals to singulars.

✓ A packet that contains an error is automatically corrected.

The use of variant plurals is becoming less common. Where once it was thought correct to base the plural form on that of the language of the root of the word, now it is almost always acceptable to use "-s" or "-es". Thus "schemata" can be "schemas", "indices" can be "indexes", and "formulae" can be "formulas". (But "radii" is not yet "radiuses" nor is "matrices" "matrixes".) Special cases remain, in particular where the plural form has replaced the singular as in "data", and in old-English forms such as "children".

Abbreviations

It is often tempting to use abbreviations such as "no.", "i.e.", "e.g.", "c.f.", and "w.r.t." These save a little space on the page, but slow readers down, particularly those whose first language is not English. It is almost always desirable to expand these abbreviations, to "number", "that is", "for example", "compared with" (or perhaps more accurately "in contrast to" since that is the sense in which "c.f." should be used), and "with respect to", or synonyms of these expressions. Where such abbreviations are used, the punctuation should be as if the expanded form were used. Also consider expanding abbreviations such as "Fig." and "Alg.", and don't use concoctions such as "1st" or "2nd". Months should not be abbreviated. Make sure that all abbreviations and acronyms are explained when they are first used.

Avoid use of "etc." A draft version of the previous paragraph began

> ✗ It is often tempting to use the abbreviations "i.e.", "e.g.", "c.f.", "w.r.t.", etc.

which is clumsier than the final form. Never write "etc., etc." or "etc. . . . "

The ellipsis is a useful notation for indicating that text has been omitted. It should, therefore, only be used in quotations.

A slash, also known as a virgule or solidus, is often used for abbreviation, as in "save time and/or space" or "used for list/tree processing". Use of slashes betrays confusion, since it is often not clear whether the intended meaning is *or* (in the usual English sense of *either but not both*), *or* (in the usual computing sense of *either or both*), *and*, or *also*. If you want to be clear, don't use slashes.

Acronyms

In technical documents with many compound terms it can be helpful to use acronyms, but as with abbreviations they can confuse the reader. An acronym is desirable if it replaces an otherwise indigestible name such as "pneumonoultramicroscopicsilicovolcanoconiosis" (miner's black lung disease), in which case the acronym becomes the name—as has happened for DNA. Frequently-used sequences of ordinary words, such as "central processing unit", are usually more convenient as acronyms; in an article about a "dynamic multiprocessing operating system", it is probably best to introduce the DMOS right at the start. However, a surfeit of acronyms will force readers to flip back and forth through the article to search for definitions. Don't introduce an acronym unless it will be used frequently.

Abbreviations are terminated by a stop but it is unusual to put stops in acronyms. Thus "CPU" is correct, "C.P.U." is acceptable but pedantic, and "CPU." is incorrect. Plurals of acronyms don't require an apostrophe; write "CPUs" rather than "CPU's".

Sexist language

Forms of expression that unnecessarily specify gender are widely regarded as sexist. In technical writing, sexist usage is easy to avoid.

✗ A user may be disconnected when he makes a mistake.

✓ A user may be disconnected when they make a mistake.

Such use of "they" as a singular pronoun is acceptable but jarring. It is preferable to recast the sentence.

✓ A user who makes a mistake may be disconnected.

Don't use ugly constructs such as "s/he", or engage in reverse sexism by using "she", unless it is absolutely impossible to avoid a generic reference. Remember that some readers find use of "he" or "his" for a generic case offensive, and dislike papers that employ such usage.

4 Punctuation

Taste and common sense are more important than any rules; you put in stops to help your readers to understand you, not to please grammarians.

Ernest Gowers
The Complete Plain Words

Punctuation is a fundamental skill. Anyone reading this book is familiar with the functions of spaces, commas, stops, capital letters, and so on. Stylistic issues of punctuation and common punctuation errors in science writing are discussed in this chapter.

Fonts and formatting

There is no obligation to use fancy typesetting just because a word processor provides it. Most computing or mathematical writing uses three fonts (plain, italic, and bold) or four (if a fixed-width font is used for the text of programs) but use of more is likely to be annoying, and all but the plain font should be used sparingly. Overuse of fonts results in messy-looking text. Some authors prefer **bold** to *italic* for emphasis, but bold print is distracting. Use of underlining for emphasis, once common because of the limitations of typewriters as typesetting devices, is obsolete.

Visual clutter of any kind is distracting and should be eliminated unless there is a clear need for it. Emphasis is one kind of clutter. Another is the use of graphic devices such as boxes around important points or icons next to results. Yet another is punctuation: excessive use of parentheses, quotes, italics, uppercase letters, and so on.

Indentation is an important tool of layout, used primarily to indicate the start of a new paragraph. Some authors prefer to use a blank line instead, a decision that is often unwise; the meaning is unclear at a page break, for example. In literature, a blank line can be used to signal the start of a new topic, a convention that however has not been adopted by the sciences. Changes of topic should be signalled by headings.

Indentation is also used to offset material that is not part of the textual flow, such as quotes, programs, and displayed mathematics. The indentation is useful because it allows easier scanning of the page.

For papers submitted for review, use wide margins and a decent font size, and don't cram lines together—referees need space for red ink. Text looks tidier if it is right-justified as well as left-justified (although it is not always easier to read). Consider using a running header, of say the authors' surnames or the paper's title, so that the paper can be reconstructed if the pages become separated. Pages should of course be numbered. Some journals request that the author information be on a separate face sheet, and many journals have specific formatting guidelines in their "Information for Authors".

Stops

Stops (or full-stops or periods) end sentences. Some writers don't use any other punctuation. Sentences should usually be short but commas and other marks give text variety. Lack of other marks makes text telegrammatic. Such text can be hard to read.

Stops are also used in abbreviations, acronyms, and ellipses. When these occur at the end of sentence, the sentence's stop is omitted. Problems with stops are a good reason to avoid abbreviations.

✗ The process required less than a second (except when the machine was heavily loaded, the network was saturated, etc.).

✓ The process required less than a second (unless, for example, the machine was heavily loaded or the network was saturated).

It is not usual to put a stop at the end of a heading.

✗ 3. Neural Nets for Image Classification.

✓ 3. Neural Nets for Image Classification

Commas

The primary uses of commas are to mark pauses, indicate the correct parsing, form lists, and indicate that a phrase is a parenthetical remark (that is, a comment) rather than a qualifier. Thus "the four processes that use the network are almost never idle" means *of the processes, the four that use the network are almost never idle*, while "the four processes, which use the network, are almost never idle" means *the four processes use the network and are almost never idle*. Incorrect use of commas in parenthetical remarks, in particular omission of the first of a pair of commas, is a frequent error.

✗ The process may be waiting for a signal, or even if processing input, may be delayed by network interrupts.

✓ The process may be waiting for a signal, or, even if processing input, may be delayed by network interrupts.

Use the minimum number of commas required for disambiguation. Sentences with many commas often have strangulated syntax; if the commas seem necessary, consider breaking the sentence into shorter ones or rewriting it altogether. But don't omit too many commas. One exception to the minimal-commas rule is to avoid ambiguity or awkward phrasing.

✗ Using disk tree algorithms were found to be particularly poor.

✓ Using disk, tree algorithms were found to be particularly poor.

Here is another example.

✗ One node was allocated for each state, but of the nine seven were not used.

✓ One node was allocated for each state, but, of the nine, seven were not used.

✓ Nine nodes were allocated, one for each state, but seven were not used.

Another exception to the minimal-commas rule is in lists. A simple example of a list is "the structures were arrays, trees, and hash tables". Many authors (and editors) prefer to omit the last comma from a list, a process that rarely adds clarity and often does it serious damage.

Commas can also be used to give the reader time to breathe.

✗ As illustrated by the techniques listed at the end of the section there are recent advances in parallel algorithms and multiprocessor hardware that indicate the possibility of optimal use of shared disk arrays by indexing algorithms such as those of interest here.

✓ As illustrated by the techniques listed at the end of the section, recent advances in parallel algorithms and multiprocessor hardware may allow optimal use of shared disk arrays by some algorithms, including indexing algorithms such as those of interest here.

Cutting this into several sentences would undoubtably improve it further.

Colons and semi-colons

Colons are used to join related statements.

✓ These small additional structures allow a large saving: costs are reduced from $O(n)$ to $O(\log n)$.

Colons are also used to introduce lists.

✓ There are three phases: accumulation of distinct symbols, construction of the tree, and the compression itself.

The elements in a list can be separated by semi-colons, allowing commas or other marks within each element.

✓ There are three phases: accumulation of distinct symbols in a hash table; construction of the tree, using a temporary array to hold the symbols for sorting; and the compression itself.

A semi-colon can also be used to divide a long sentence, or to set off part of a sentence for emphasis.

✓ In theory the algorithm would be more efficient with an array; but in practice a tree is preferable.

Colons and semi-colons are valuable but should not be overused.

Apostrophes

Many people seem to have trouble with apostrophes; even professional writers get them wrong now and again. But the rules are quite simple.

- Singular possessives such as "the student's algorithm", "Gowers's book", and "Su and Ling's method" require an apostrophe and an "s".
- Plural possessives such as "students' passwords" require an apostrophe but no "s".
- Pronoun possessives such as "its" (as in "its speed") and "hers" do not require an apostrophe.
- Contractions such as "it's" (as in "it is blue") and "can't" require an apostrophe; but note that contractions should be avoided in technical writing.

Other than in the cases above, apostrophes are not required.

Exclamations

Avoid exclamation marks! Never use more than one!!

The proper place for an exclamation mark is after an exclamation (such as "Oh!"—not a common expression in technical writing), or, rarely, after a genuine surprise.

✓ Performance deteriorated after addition of resources!

This is acceptable but not particularly desirable. It would be better to omit the exclamation and add emphasis some other way.

✓ Remarkably, performance deteriorated after addition of resources.

Hyphenation

Many compound words, such as "database", would originally have been written as two separate words, "data base". When the combination becomes common, it is hyphenated, "data-base", then eventually the hyphen is dropped to give the final form. Some words are in a state of transition from one form to another. In the database literature, for example, all three of "bit slice", "bit-slice", and "bitslice" are used with regard to signature files, and in general writing both "co-ordinate" and "coordinate" are common. Make sure that you are consistent.

Hyphens are also used to override right associativity. We parse phrases such as "randomized data structure" into *randomized data-structure*, and thus realize that the topic is not a structure for randomized data. In some phrases that are not right-associative, such as "skew-data hashing", we need the hyphen to disambiguate (although in this case it might be better to write "hashing for skew data").[9] Sometimes there is no correct hyphenation and the sentence has to be rewritten. The phrase "hash based data structure" should be written "hash-based data structure", but "binary tree based data structure" should probably be written, albeit awkwardly, as "data structure based on binary trees".

Good word-processors hyphenate words when they run over the end of a line, to preserve right-justification. Automatic hyphenation is not always correct and should be checked, to ensure that syllables are not broken or that the break is not too close to the end of the word. For example, the hyphenations "mac-hine" and "availab-le" should be corrected (to "mach-ine" and "avail-able"), and "edited" should probably not be hyphenated at all.

Note that there are three distinct "dash" symbols: the hyphen "-" used for joining words, the minus sign or en-dash "–" used in arithmetic and for ranges such as "pages 101–127", and the em-dash "—" used for punctuation.

Capitalization

Capital letters were once used far more liberally than they are now: in the eighteenth century writers commonly used capitalization (that is, an initial capital letter) to denote nouns. Today, only proper names are capitalized, and even these can be in lower case if the name is in common use; for example, the capitals in the phrase "the Extensible Hashing method" should be converted to lower case.

Some names are not consistently capitalized, particularly those of programming languages. Acronyms that cannot be sounded, such as "APL", should always be written that way, but the only general rule for other cases is to follow other authors. For example, consider the names

[9] There is a hyphen missing in the headline "Squad helps dog bite victim".

"FORTRAN" and "Prolog", both of which are abbreviations derived from truncated words. These are however proper names and should always have an initial capital; "lisp" and "pascal" are incorrect.

In technical writing it is usual to capitalize names like "Theorem 3.1", "Figure 4", and "Section 11". In other writing, lower case is preferred, but in technical writing lower case looks sloppy to some readers.

Headings can be either minimally or maximally capitalized. In the former, words are capitalized as they would be in normal text, except that the word following a colon is capitalized.

✓ The use of jump statements: Advice for Prolog programmers

In the latter, words other than articles, conjunctions, or prepositions are capitalized; even these may be capitalized if they are over three letters long.

✓ The Use of Jump Statements: Advice for Prolog Programmers

The same rules apply to captions and titles of references.

Be consistent in your style of capitalization. It is acceptable to use maximum capitalization for sections and minimum capitalization for subsections, but not the other way around.

Quotations

The convention for punctuation of quotations has been to put commas and stops inside the quotation marks even when they are not part of the original material. However, outside the United States this convention is in decline. A better convention is to place a punctuation symbol within the quotation marks only if it was used in the original text—such as when a complete sentence is being quoted—as is done throughout this book.

✓ Crosley [14] argues that "open sets are of insufficient power", but Davies [22] disagrees: "If a concept is interesting, open sets can express it."

(But note that it is not necessary to quote such a dull statement as "open sets are of insufficient power"; paraphrase, or even simply omitting the quote symbols, would be more appropriate. Omission of quotation marks in this case is acceptable—that is, not plagiarism—because Crosley's statement is a natural way to express the concept.)

If the material in the quotation marks is a literal string, the punctuation must go outside. Since most punctuation symbols have meaning in programming languages, when programming statements are quoted the matter in the quote will be syntactically incorrect if the punctuation is in the wrong place.

× One of the reserved words in C is "for."

✓ One of the reserved words in C is "for".

Some editors will change this to the wrong form. You may prefer to avoid the problem altogether.

✓ One of the reserved words in C is `for`.

Note that quotation symbols (" and ") are not the same as the ASCII double-quote symbol (`"`).

Parentheses

A sentence containing a statement in parentheses should be punctuated exactly as if the statement was removed; and the punctuation of the parenthetical statement should be independent of the rest of the sentence.

× Most quantities are small (but there are exceptions.)

✓ Most quantities are small (but there are exceptions).

× (Note that outlying points have been omitted).

✓ (Note that outlying points have been omitted.)

Parenthetical remarks should be asides that the reader can ignore—important text should not be in parentheses. The same rule applies to footnotes; if you think that some text should be relegated to a footnote then perhaps it can be deleted.

Overuse of parentheses looks crowded; avoid having more than one parenthesized remark per paragraph, or more than a couple per page. Parentheses within parentheses are hard to read and look like an editing error. Get rid of them.

The use of "(s)" to denote the possibility of a plural, as in "any observed error(s)", is ugly and almost never necessary; it is better to omit the parentheses or recast the sentence.

Citations

Citations should be punctuated as if they were parenthetical remarks.

✗ In [2] such cases are shown to be rare.
  In (Wilson 1984) such cases are shown to be rare.

Some journals typeset citation numbers as superscripts, in which case this example becomes "In[2] such cases are shown to be rare". Never treat a bracketed expression, whether a citation or otherwise, as if a word.

✓ Such cases have been shown to be rare [2].
  Such cases have been shown to be rare (Wilson 1984).
  Wilson [2] has shown that such cases are rare.
  Wilson has shown that such cases are rare [2].
  Wilson (1984) has shown that such cases are rare.

The cite should be close to the material it relates to—poor placement of cites can be ambiguous.

✗ The original algorithm has asymptotic complexity $O(n^2)$ but low memory usage, so it is not entirely superseded by Ahlberg's approach, which although of complexity $O(n \log n)$ requires a large in-memory array [7,19].

Since Ahlberg did not recognize the array as a problem and does not describe the old approach, this sentence is misleading.

✓ The original algorithm has asymptotic complexity $O(n^2)$ but low memory usage [19], so it is not entirely superseded by Ahlberg's approach [7], which although of complexity $O(n \log n)$ requires a large in-memory array.

The placement of citations depends partly, however, on the citation style used. With the superscript style, for example, it is usual to try and place citations at the end of the sentence.

Formatting of citations and bibliographies is described in detail in Mary-Claire van Leunen's *Handbook for Scholars* and in *The Chicago Manual of Style*.

5 Mathematics

Mathematics is no more than a symbolism. But it is the only symbolism invented by the human mind which steadfastly resists the constant attempts of the mind to shift and smudge the meaning ... Our confidence in any science is roughly proportional to the amount of mathematics it employs—that is, to its ability to formulate its concepts with enough precision to allow them to be handled mathematically.

J. Bronowski and Bruce Mazlish
The Western Intellectual Tradition

Mathematics gives solidity to abstract concepts. As for English in general, there are conventions of presentation for mathematics and mathematical concepts. Reading mathematics is hard work at the best of times, unpleasant work if the mathematics is badly presented. In this chapter I give general guidelines for mathematical style. Other guides are Leonard Gillman's *Writing Mathematics Well* [14] and Nicholas J. Higham's *Handbook of Writing for the Mathematical Sciences* [15].

Clarity

In mathematical writing it is essential to be precise. For example, an ambiguous statement of a theorem can make its proof incomprehensible. This rule particularly applies to fundamental definitions, as discussed on page 8. Many terms have well-defined mathematical meanings and are confusing if used in another way.

Normal, usual. The word "normal" has several mathematical meanings; it is often best to use, say, "usual" if a non-mathematical meaning is intended.

Definite, strict, proper, all, some. Avoid "definite", "strict", and "proper" in their non-mathematical meanings, and be careful with "all" and "some".

Intractable. Formally, an algorithm or problem is "intractable" only if it is NP-hard, that is, has computational complexity that is worse than polynomial. "Intractable" is sometimes used to mean *hard to do*, which is acceptable as long as there is no possibility of confusion.

Formula, equation. A "formula" is not necessarily an "equation"; the latter involves an equality.

Equivalent, similar. Two things are "equivalent" if they are indistinguishable with regard to some criteria. If they are not indistinguishable, they are at best "similar".

Element, partition. An "element" is a member of a set (or list or array) and should not be used to refer to a subpart of an expression. If a set is "partitioned" into subsets, the subsets are disjoint and form the original set under union.

Average, mean. "Average" is used loosely to mean *typical*. Only use it in the formal sense—of *arithmetic mean*—if it is clear to the reader that the formal sense is intended. Otherwise use "mean" or even "arithmetic mean".

Subset, strict. "Subset" should not be used to mean *subproblem*. Orderings (or partial orderings) specified in English are assumed to be non-strict. For example, "A is a subset of B" means that $A \subseteq B$; to specify $A \subset B$ use "A is a strict subset of B". The same rule applies to "less than", "greater than", and "monotonic".

Theorems

Many readers skim articles to find theorems (or other results such as illustrations or tables). For this reason, and because they may be quoted verbatim in other articles, theorems should as far as possible be independent of the rest of the text.

Definitions, theorems, lemmas, and propositions should be numbered, even if there are only two or three of each in the paper, and you should also consider numbering examples. Not only does numbering allow reference within the paper, but it simplifies discussion of the paper later on: it is much easier for a correspondent to refer to "definition 4.2" than "the definition towards the bottom of page 6".

Some presentation problems are intractable. For a theorem with a complex proof, if the lemmas are proved early they appear irrelevant, and if they are proved late the main proof is harder to understand. One solution is to state the main theorem first then state and prove the lemmas before giving the main proof, but in other cases all that can be done is to take extra care in the motivation and make liberal use of examples. Explain the structure of long proofs before getting to the detail, and explain how each part of the proof relates to the structure.

When you publish a paper containing a proof of a theorem, always be satisfied that the proof is correct. However, the details of the proof may not be important to the reader and can often be omitted. Steps in the logic of a proof should be simple enough that the gaps can be completed by a reader mechanically, without too much invention. A common mistake is to unnecessarily include mechanical algebraic transformations; you need to work through these to check the proof, but the reader is unlikely to find them valuable. When presenting your proof—that is, making it comprehensible to a reader—remember that you are presenting a reasoned argument. Use any available means to convey your argument with the greatest possible clarity; a diagram, for example, is perfectly acceptable.

Proof by contradiction is overused. By all means use contradiction to develop your understanding of the problem, if it helps you to get the details right, but it is always worthwhile exploring how to achieve the result directly instead of by contradiction.

It helps the reader if the end of each proof, example, or definition is marked with a symbol such as a box. Alternatively, proofs and so on can be indented to set them apart from the running text.

Readability

Mathematics is usually presented in italics, to distinguish it from other text. Thus in the expression "of length n" it is clear that n is a variable of some kind. The main exception is function names, such as log or sin, which are written in an upright font. The pairs "[ ]" (that is, brackets), "( )" (parentheses), and "{ }" (braces) are all used to delimit subexpressions, but braces can be confusing because they are also used to denote sets. Use parentheses of appropriate size; they should, more or less, be a little taller than the expressions they enclose.

✗ $(p \cdot (\sum_{i=0}^{n} A_i))$

✓ $\big(p \cdot (\sum_{i=0}^{n} A_i)\big)$

Sentences with embedded mathematics should be structured as if each formula was a simple phrase. Phrases indicate how the following text will be structured, but formulas do not, and so should not be used at the start of a sentence.

✗ $p \leftarrow q_1 \wedge \cdots \wedge q_n$ is a conditional dependency.

✓ The dependency $p \leftarrow q_1 \wedge \cdots \wedge q_n$ is conditional.

Give the type of each variable every time it is used, so that the reader doesn't have to remember as many details and can concentrate on content, and watch out for misplaced types or variables.

✗ The values are represented as a list of numbers L.

✓ The values are represented as a list L of numbers.

The former version is ambiguous—the symbol L might denote an individual member of the set.

Formulas embedded in text should not run together.

✗ For each x_i, $1 \le i \le n$, x_i is positive.

✓ Each x_i, where $1 \le i \le n$, is positive.

Mathematics should not take the place of text: readers will quickly get lost if they must decipher a stream of complex expressions.

✗ Let $\langle S \rangle = \left\{ \sum_{i=1}^{n} \alpha_i x_i \mid \alpha_i \in F, 1 \le i \le n \right\}$. For $x = \sum_{i=1}^{n} \alpha_i x_i$ and $y = \sum_{i=1}^{n} \beta_i x_i$, so that $x, y \in \langle S \rangle$, we have $\alpha x + \beta y = \alpha \left(\sum_{i=1}^{n} \alpha_i x_i \right) + \beta \left(\sum_{i=1}^{n} \beta_i x_i \right) = \sum_{i=1}^{n} \left(\alpha \alpha_i + \beta \beta_i \right) x_i \in \langle S \rangle$.

Although the mathematics in this example is fairly clear, there is no motivation, and the thicket of symbols is daunting.

✓ Let $\langle S \rangle$ be a vector space defined by

$$\langle S \rangle = \left\{ \sum_{i=1}^{n} \alpha_i x_i \mid \alpha_i \in F, 1 \le i \le n \right\} .$$

We now show that $\langle S \rangle$ is closed under addition. Consider any two vectors $x, y \in \langle S \rangle$. Then $x = \sum_{i=1}^{n} \alpha_i x_i$ and $y = \sum_{i=1}^{n} \beta_i x_i$. For any constants $\alpha, \beta \in F$, we have

$$\begin{aligned} \alpha x + \beta y &= \alpha \left(\textstyle\sum_{i=1}^{n} \alpha_i x_i \right) + \beta \left(\textstyle\sum_{i=1}^{n} \beta_i x_i \right) \\ &= \sum_{i=1}^{n} \left(\alpha \alpha_i + \beta \beta_i \right) x_i \, , \end{aligned}$$

so that $\alpha x + \beta y \in \langle S \rangle$.

Note the vertical alignment of the equality symbols.

Important or complex formulas should be displayed. In such displays, the formula can be either centred or indented; choose either, but be consistent. However, if part of the display is an algorithm or program, centering can look peculiar. Displayed formulas (or graphs or diagrams) should be positive results, not counter-examples, so that readers who skim through the paper won't be misled. If a displayed formula is sufficiently important it should be numbered, to allow discussion of it elsewhere in the paper and for reference once the paper is published. As in the example above a displayed formula, like an embedded formula, should be treated as a phrase.

Mathematical symbols should, if possible, be the same font size as other characters. For example, the expression $(n(n + 1) + 1)/2$ is more legible than $\frac{n(n+1)+1}{2}$ even though the former uses more characters; but take care to avoid potentially ambiguous expressions such as $a/b + c$. Consider breaking down expressions to make them more readable, especially if doing so enlarges small symbols.

$$\times \quad f(x) = e^{2^{-\frac{b}{a}x\sqrt{1-\frac{a^2}{x^2}}}}$$

$$\checkmark \quad f(x) = e^{2^{g(x)}} \quad \text{where} \quad g(x) = -\tfrac{b}{a}x\sqrt{1-\tfrac{a^2}{x^2}}$$

Avoid unnecessary subscripts: use x and y rather than x_1 and x_2. Also, don't pile subscripts on top of each other: the symbol i is legible in x_i, barely acceptable in x_{j_i}, and ridiculous in $x_{k_{j_i}}$. Mix subscripts and superscripts with caution: the expression $x_{j'}^{p_k}$ is a mess. Be careful with choice of letters for subscripts: in some small fonts, the letters i, j, and l are not easy to distinguish. Likewise, sets and set notation can be used to simplify presentation of mathematics and algorithms. For example, $\sum_{i=1}^{k} f_{w_i}$ is harder to read than $\sum_{w \in W} f_w$.

Notation

Ensure that the symbols you use will be correctly understood by, and familiar to, the reader. For example, there are several symbols ($\Rightarrow$, $\mapsto$, $\vdash$, $\supset$, $\sqsupset$, $\models$, and probably others) that are used in one context or another for logical implication. These symbols also have other meanings, so there is plenty of scope for confusion. The symbols $\sim$, $\simeq$, and $\approx$ are all used to mean *approximately equal to*, but $\sim$ is also used in other contexts. The symbol $\cong$ means *is congruent to*, not *approximately equal*.

Symbols such as $\forall$, $\exists$, $<$, $>$, $=$, and $\Rightarrow$, and abbreviations such as "iff", should not be used as substitutes for words. These symbols may be compact but they are hard for readers to digest. But don't replace symbols by words unnecessarily; for example, write "$a \leq b$" rather than "a is less than or equal to b". Concocted or amusing symbols are not a good idea; don't use $\clubsuit$ or $\natural$ as operators, for example.

Don't reuse notation: an excellent way of confusing readers is to use N for one quantity on page 6 and for another on page 13. Likewise, use consistent notation for expressions with similar meaning. Develop and adhere to conventions such as using i and j for integer subscripts and uppercase letters for sets. Don't vary an existing notation without good reason.

Take care with accents. Don't use $\hat{a}$, $\tilde{a}$, $\bar{a}$, and $\vec{a}$ together, and don't pile up primes: the symbol a'' may be clear, but what about $D_{i'}^{l''}$? Some authors put powers on primes, as in a'^4 to represent a'''', but this notation is often unclear. If you have such deep primes consider reworking your notation to get rid of them.

Ranges and sequences

The *closed* range of real numbers r where $a \leq r \leq b$ is represented by "$[a, b]$", the *open* range $a < r < b$ is represented by "(a, b)", the range $a \leq r < b$ is represented by "$[a, b)$", and the range $a < r \leq b$ is represented by "$(a, b]$".

It is common practice to use an ellipsis to describe a sequence of integers; thus $m, \ldots, n$ is all integers between m and n inclusive. An infinite sequence is usually represented by $m_1, m_2, \ldots$, where it is assumed that the reader can extrapolate from the initial values to the other members of the sequence. Thus "$2, 4, 8, \ldots$" would be assumed to be the sequence of positive powers of 2. Always state both the lower and the upper bound if the sequence is finite and ensure that the intended sequence is clear.

An expression such as $1 \leq i \leq 6$ should be replaced by $i = 1, \ldots, 6$ if it is not clear that i should be an integer.

Alphabets

Use of characters from the Greek alphabet to denote variables and quantities can add clarity to mathematical writing, because these characters cannot form English words and so cannot be confused with the text. But they should not be overused.

Many readers are familiar with only a few of the Greek letters and use of unfamiliar letters should be minimized, if only because use of any new notation should be minimized. Most people find it easier to remember that a letter denotes a certain quantity if they know the name of the letter; if they do not know the name they tend to invent one, but this invention is generally not as effective a label as a real name. For example, reading the statement "sets are denoted by α" might result in the thought *sets are denoted by alpha* while reading "sets are denoted by ϱ" (a form of rho) might result in the thought *sets are denoted by squiggle-that-looks-like-a-bit-like-a-backwards-g*. Other characters that have this effect are the Greek letters ζ (zeta), ξ and Ξ (xi), and symbols such as $\aleph$, $\Re$, and $\Im$.

Some mathematical symbols and characters from other alphabets have a superficial resemblance to more familiar symbols. Some pairs that can cause confusion, particularly after imperfect reproduction, are shown below.

Symbol		Confused with	
ϵ	epsilon	e	
η	eta	n	
ι	iota	i	
μ	mu	u	
ρ	rho	p	
υ	upsilon	v	
ω	omega	w	
$\vee$	or	v	
$\propto$	proportional	α	alpha
$\emptyset$	empty set	ϕ	phi

Never use handwritten symbols in a submitted paper. If you can't print the symbol you want to use, change it.

Line breaks

Avoid letting a number, symbol, or abbreviation appear at the start of the line, particularly if it is the end of a sentence.

✗ We have therefore used an additional variable, denoted by
x. It allows ...

✓ We have therefore used an additional variable, denoted by x.
It allows ...

✗ Accesses to the new disk can be performed in about 12
ms using our techniques.

✓ Accesses to the new disk can be performed in about 12 ms
using our techniques.

Most word processors provide an unbreakable-space character that prevents this behaviour. However, some word processors insist on breaking lines at awkward places in mathematical expressions.

✗ The problem can be simplified by using the term $f(x_1, \ldots, x_n)$ as a descriptor.

Sometimes the only solution is to rewrite the surrounding text.

✓ The problem is simplified if the term $f(x_1, \ldots, x_n)$ is used as a descriptor.

Numbers

In technical writing, numbers should usually be written as figures, not spelt out. The common exceptions are: approximate numbers; numbers up to twenty, unless they are literal values or part of an expression of measurement; and numbers at the start of a sentence, although it is generally better to recast the sentence so that the number is elsewhere. Percentages should always be in figures.

✗ 1024 computers were linked into the ring.
Partial compilation gave a 4-fold improvement.
The increase was over five per cent.
The method requires 2 passes.

✓ There were 1024 computers linked into the ring.
Partial compilation gave a four-fold improvement.
The increase was over 5 per cent.
The increase was over 5%.
Method 2 is illustrated in Figure 1.
The leftmost 2 in the sequence was changed to a 1.
The method requires two passes.

Don't mix modes.

✗ There were between four and 32 processors in each machine.

✓ There were between 4 and 32 processors in each machine.

In English-speaking countries, the traditional method for separating long sequences of digits into groups of three has been to use commas, as in "1,897,600". This method has two disadvantages: it can be ambiguous if the numbers are part of a comma-separated list, and decimal points are denoted by commas in many countries, so that a number such as "1,375" could be misinterpreted. It is for these reasons that the alternative of using thin spaces has been introduced, as in "1 897 600". But the comma-separated style remains popular in many countries.

Fractions are only rarely used for values, and should never be used as abbreviations.

✗ About 1/3 of the data was noise.

✓ About one-third of the data was noise.

As for mathematical symbols in general, numbers should not be used to start a sentence. Nor should they be adjacent.

✗ There were 14 512-Kb sets.

✓ There were fourteen 512-Kb sets.

Never omit the leading 0 in numbers whose magnitude is less than 1; write "the size was 0.3 Kb", not "the size was .3 Kb".

Avoid the phrase "orders of magnitude".

✗ The new algorithm is at least two orders of magnitude faster.

In this example, is the unit of magnitude binary or decimal? It would be better to be explicit.

✓ The new algorithm is at least a hundred times faster.

Numbers of the same units should for consistency be represented to the same precision. In physical experiments, it is usual to represent numbers to the same relative precision, that is, the same number of digits. In computer science, in which values are usually measured to the same absolute precision, it is more logical to represent numbers to the same units.

✗ The sizes were 7.31 Kb and 181 Kb respectively.

✓ The sizes were 7.3 Kb and 181.4 Kb respectively.

In a published paper the same figure was quoted in different places as "almost 200 000", "about 170 000", "173 000", and "173 255"—an entirely unnecessary inconsistency.

Be realistic about accuracy and error. Your system may report that a process required 13.271844 CPU seconds, but in all likelihood the last four or five digits are meaningless. You should not imply accuracy by including spurious numbers. For example, "0.5 second" is not equivalent to "half a second", since the former implies that careful measurements were taken. Guesses and approximations should be clearly indicated as such, with words such as "roughly", "nearly", "approximately", "almost", or "over"; but don't use wordy phrases such as "in the region of".

Percentages

Use percentages with caution.

✗ The error rate grew by 4%.

This example is ambiguous because an error rate is presumably a percentage. It is better to be explicit, and to avoid mixing kinds of percentages.

✗ The error rate grew by 4%, from 52% to 54%.

✓ The error rate grew by 2%, from 52% to 54%.

When stating a percentage, ensure that the reader knows what is a percentage of what. If you write that "the capacity decreased by 30%", is this 30% of the old figure or the new? The convention is to use 100% as the starting point, but in a series of statements of percentages it is easy to get lost. Use percentages rather than odds to express probabilities.

✗ The likelihood of failure is 2:1.

✓ The likelihood of failure is one in three.

✓ The likelihood of failure is about 30%.

Don't use probabilities to describe small sets of observations. Success in two of five cases does not mean that the method "works 40% of the time". The percentage gives the result authority it does not deserve.

Units of measurement

Two quantities are commonly measured in computer science: space and time. For time, the basic units are the second (sec), minute (min) and hour (hr); note that it is unusual to give the abbreviated forms of these units. For the divisions of the second—the millisecond (ms or msec), microsecond (μs or μsec), and nanosecond (ns or nsec)—some readers may be unsure of the notation. For example, "ms" might be interpreted as *microsecond*. State such units in unabbreviated form at least once.

When writing about hours or minutes use a colon rather than a stop to separate the components of the time. That is, write "3:30 minutes" rather than "3.30 minutes".

For space the basic units are bit and byte. These are combined in tenth powers of 2 rather than third powers of 10.

Unit	value (bytes)	denotation
kilobyte	$2^{10} \approx 10^3$	Kb, Kbyte
megabyte	$2^{20} \approx 10^6$	Mb, Mbyte
gigabyte	$2^{30} \approx 10^9$	Gb, Gbyte
terabyte	$2^{40} \approx 10^{12}$	Tb, Tbyte
petabyte	$2^{50} \approx 10^{15}$	Pb, Pbyte
exabyte	$2^{60} \approx 10^{18}$	Eb, Ebyte

If there is any likelihood that, for example, a reader could interpret "Mb" as *megabit*, use "Mbyte" or "megabyte" instead. The larger units, especially "Tb", "Pb", and "Eb", are unfamiliar to most readers and should be written in full at least once, preferably with an explanation.

There are few derived units in computing other than the transfer rate of bytes per second, as in "18 Mb/sec". I was most surprised to encounter "millibits" in a paper on arithmetic coding (in which symbols can be represented in a fraction of a bit). The unit is so unusual that "thousandths of a bit" would be preferable.

Choose units that are easy to understand. For example, seconds can be preferable to minutes because fractions of a minute can be confusing: does "1.50 minutes" mean *one and a half minutes* or *one minute and fifty seconds*? (This problem can be avoided by using colons instead of stops, as discussed above.) Also, as values such as clock speeds and transfer rates are quoted in seconds, use of minutes makes comparison more difficult. On the other hand, "13:21 hours" is probably kinder to the reader than "47.8×10^3 seconds".

Some units, although in general use, are not well-defined. For example, MIPS (million instructions per second) is almost meaningless as it cannot be used to compare machines of different architectures.

For quantities greater than 1, the unit is plural. For smaller quantities, the unit is singular.

✓ The average run took 1.3 seconds, and the fastest took 0.8 second.

Units should be typeset in the font used in the paper for text, even when they are part of a mathematical expression.

✗ The volume is r^p Kb in total.

✓ The volume is r^p Kb in total.

Put white space between values and units. Write "11.2 Kbytes", not "11.2Kbytes". Numbers and their units should be hyphenated when they are used as an adjective.

✓ We also tried the method on the 2.7-Kb input.

6 Graphs, figures, and tables

"And what is the use of a book", thought Alice, "without pictures or conversations?"

Lewis Carroll
Alice in Wonderland

Some information is best presented in a pictorial form, such as a graph or figure, to show trends and relationships. Other information is best as a table, to show regularities. In this chapter I consider style issues related to such illustrations. A good text on construction and choice of graphs is Edward R. Tufte's *The Visual Display of Quantitative Information* [20].

Illustrations

Well-chosen illustrations breathe life into a paper, giving the reader interesting visual elements to browse and highlighting the central results and ideas. A typical figure is of visual matter such as a graph or diagram, or of textual matter such as a table, algorithm, or, less commonly, complex mathematics. A figure is usually at the top or the bottom of a page, or on a page by itself, to set it apart from ordinary text. Because illustrations attract the attention of readers they should only be used for matter that is central to the paper.

Each figure should be numbered to allow easy reference and have a descriptive caption so that the figure is, as far as possible, independent of the text. An illustration should always be introduced and discussed, preferably just before or on the page on which it occurs. If you don't have anything to say about an illustration, leave it out.

Illustrations are covered by copyright; figures from another source can only be reused with permission of the author and the publisher of the original. If you reuse a figure, get permission to do so and identify the original author and source, preferably in the caption. You may also need to include the original copyright statement.

Graphs

Graphs are usually the best way to present numerical results. Numbers should be used sparingly; instead, use graphs wherever appropriate. Summarize numbers with a graph in the main text, so that the behaviour under discussion is obvious. If you must list the numbers as well put a detailed table of results in an appendix, but in many cases the numbers are only of transient significance and can be omitted.

Avoid flooding the document with statistics, even in graphical form—three or fours graphs should be enough, and ten is almost certainly too many. It is all too easy to generate reams of numbers by running software with different combinations of parameters, but, even though these numbers may contribute to your analysis and understanding of the phenomena being observed, they are unlikely to be of value to a reader. The information you present should be selected because it is supporting evidence for a hypothesis, not because it is an output of some program.

Graphs should be kept simple, with no more than a few plotted lines and a minimum of clutter. The horizontal or x-axis should be used for the parameter being varied, or the input; the vertical or y-axis is for the function of the parameter, or the output. Plotted lines of discrete data should always have points marked, as in the graphs at the end of this chapter, by distinctive marks such as circles, boxes, or triangles. Don't use ticks or crosses as these are harder to see. The lines, axes, and other elements should be of similar thicknesses—don't mix a large, bold font with lightly-drawn lines, for example. Secondary marks, such as axis ticks, are usually a little lighter than the other elements. If you use shades of grey to distinguish different elements in the graph, ensure that the shades are sufficiently distinct; also, lines in lighter grey sometimes need to be a little thicker than other lines.

You may need a little imagination to allow the desired picture to emerge. Logarithmic axes are useful because they show behaviour at different orders of magnitude. An example of changing to a logarithmic axis is shown on page 89. Graphs with logarithmic axes are also useful when plotting problem size against algorithm running time, as different asymptotic growth rates give straight lines of different slope. If the relationship is more complex, some sort of transformation on the data may yield a straight line or some other simple curve.

In some cases data that seems innately tabular can be represented as a graph. Often a bar graph is suitable because the items being compared are not ordered; the graph on page 90 is an example. (Such data should not be represented by joined points, which would imply that the axes were related by a function.) Another example of how to represent tabular data as a graph, for the more complex problem of comparing space and time simultaneously, is the graph on page 91.

Graphs are used to illustrate change in one parameter as another is varied. In some cases more than two parameters can interact in complex ways. If two parameters, say A and B, depend on a third, C, then a good solution is to plot C on the x-axis and have two y axes, one for each of A and B, as illustrated in the graph on page 88. If two parameters, say D and E, jointly determine a third, F, in some complex way—thus describing a three-dimensional space—the problem is more difficult. The best solution is to experimentally graph D against F for several fixed values of E, and use these results to choose an E value that yields a representative graph; and similarly vary E for several fixed D, to choose a representative D.

Where different methods of achieving the same aim are being illustrated, for example comparing the suitability of different data structures for some purpose, and a separate graph is used for each method, then the axes in each graph should have the same scale. That is, if y, say, ranges from 0 to 80 on one graph, it should also range from 0 to 80 on the other, to allow direct comparison between the methods. Comparison is easier with several (but not too many) lines on one graph.

Beware of using graphs to make unsupported claims. For example, consider the "space wastage" line in the graph on page 88: it would not be possible to identify the slope of this line with any confidence, nor identify it as a particular kind of curve. The only reasonable inference would be that increasing list length increases space wastage.

There are several good software packages for drawing graphs. Valuable features include: ability to place several lines on one graph; an assortment of symbols (such as crosses, squares, and triangles) for marking points; optional connection of points with solid, dotted, or dashed lines, and optional omission of the point marks; multiple font sizes and line thicknesses; availability of greys and colours; optional logarithmic or exponential scaling on both axes; axis editing, to specify where the ticks are placed, how many digits of precision to use, and what range to cover; ability to move and rotate the legend or key, any line labels, the axis labels, and the graph label; and ability to apply simple functions to (x, y) values. Most of these features were used in the example graphs at the end of this chapter.

Diagrams

Diagrams (or schematics) are put to many uses in papers about computing. They illustrate processes or architectures; explain data structures and algorithms; present relationships; and show examples of interfaces. There are areas of computer science in which the diagrams are, in some sense, the result being presented in the paper: entity-relationship models are diagrams conforming to a well-defined notation, for example, and automata are often described by diagrams. Many areas of research have highly developed conventions and standards for diagrams; browsing a few relevant papers from an area usually gives a good idea of what elements a diagram should incorporate and of how it should be presented.

Broadly speaking, diagrams are used to show either a structure, a process, or a state. Although these are rather high-level distinctions, they are valuable because a common mistake in design of diagrams is to attempt to combine these purposes inappropriately. For example, a schematic showing data flow in an architecture is likely to be unclear if control flow is also illustrated.

Use preliminary hand sketches to develop the diagram. This early stage is the appropriate time to balance the diagram, by checking that it is well-proportioned (half as wide again as it is high is about right); makes good use of the space; is laid out well and doesn't have the elements bunched to one side; and the relative sizes of the elements look reasonable. However, never submit a paper with a hand-drawn diagram unless it has been prepared by a professional; almost any diagram can be drawn well with the tools available on a basic computer.

A diagram should not be too dark; keep it as open as possible. This is best achieved by eliminating all clutter. A diagram does not have to be too faithful to every detail of the concept being illustrated; fine details can always be clarified in the supporting text and even the best diagram requires some explanation. Use meaningful labels, which should be displayed horizontally, and make the point size and font of the labels similar to that of the other text. As for text in general, there should be no more than two or three fonts and font sizes.

Lines should not be too heavy—at most a little thicker than the lines used to draw the text font. Shades of grey can be used to distinguish between solids but are not as effective for distinguishing between lines, and don't use shades that are too light or too similar. Pictorial elements should be used consistently, so that, for example, arrows and lines of the same kind have the same meaning. If arrows are used to show arcs as well as to point at features, distinguish them by, say, using dashed lines in one case and solid lines in another.

Diagrams, like graphs, can add greatly to the clarity of a paper. But be aware that the design of good diagrams is an art—expect to revise your pictures as often as you would your text. A weak diagram is shown on page 155 and a revision of it is on page 156. Other diagrams are shown on pages 94, 95, and 101.

Tables

Tables are useful for presentation of information that is unsuitable for graphs or figures, such as the properties of each of a series of datasets or data where the exact values are important. The tables on pages 92 and 93 have appropriate content, although the first is poorly laid out.

A well-designed table is hierarchical. Simple tables are an arrangement of columns and rows, in which each column has a heading at the top and each row has a label or stub at the left. In more complex tables, columns and rows may be partitioned. The hierarchy can be indicated in several ways: rows or columns can be separated by double lines, single lines, or white space; headings can span several columns; labels can refer to several rows. Deeper structure—sometimes necessary but usually unwise—can be indicated by markup within the table such as embedded headings. (A complex table is shown on page 93.) The items below a column head should be of the same kind or about the same thing. Items to the right of a row label should all be properties of the label. The column of labels does not need to have a heading, but this position, the top-left corner of the table, should not be a label for the other column headings. If there is no heading for the column of labels, leave the position blank.

Don't have too many horizontal or vertical rules. In particular, there is no need to have a rule between every row or column. (An example of this error is shown on page 92.) But do have rules between groups of rows, and, in rare cases, between groups of columns, to act as guides and to separate items that don't belong together. Don't make tables too dense: rather than cram in a large number of columns, have two tables, or, even better, be selective about the information you present. In most tables no position should be blank; if there is no applicable value, put in a dash, and explain somewhere what it means. Values of the same units in a column should be aligned in a logical way. Numbers should be aligned on the decimal point.

Using tables to show function values at different points is usually not a good idea because graphs serve this purpose well; a possible exception is when a function only has two or three values, in which case a graph would be too simple or sparse to be of interest. In some cases, such as a table or graph that does no more than illustrate a simple relationship, consider stating the relationship and omitting the illustration altogether.

✗ As illustrated in Table 6, temporary space requirements were 60% to 65% of the data size.

✓ In our experiments, temporary space requirements were 60% to 65% of the data size.

Small tables can be part of the running text, displayed in the same way as mathematics. Larger tables should be labelled and positioned at the top or bottom of a page.

Understanding a table of any complexity is hard work. For presentation of results, graphs or explanatory text are preferable; have a table to which the interested reader can refer, but don't rely on a table to convey essential information.

Axes, labels, and headings

The space constraints on axes, labels, and headings mean that some terms have to be abbreviated; for example, see the table on page 157. It is helpful to state these terms in full in the text discussing the illustration, but do so in a natural way.

✗ The abbreviations "comp.", "doc.", and "map." stand for "compression", "document", and "mapping table" respectively.

✓ The effect of compression on the documents and the mapping table is illustrated in the second and third rows.

Where appropriate, units should be stated in labels. Write "Size (bytes)", not just "Size".

Some readers get confused by scaling on axes and labels. Suppose for example that an axis is labelled as "CPU time (seconds $\times 10^{-2}$)". The convention is that the reader should multiply axis values by 10^{-2}, so that 50 means 0.5. But some readers may assume that the axis values have already been multiplied by 10^{-2}, so that they read 50 as 5000. Mentioning typical values in the text discussing the illustration will avoid this problem. In the case of a graph it is helpful anyway to include some representative numbers in the text, because graphs are hard to read with any precision.

✓ Figure 4 shows how time and space trade off as node size is varied; as can be seen, response of under a second is only possible when size exceeds 11 Kb.

Captions and labels

Captions and labels should be informative. It is conventional in computer science for captions to be only a few words, but I prefer captions that are more descriptive, with explanation of the figure's major elements. (A diagram and caption are discussed on page 94.) Use either minimum or maximum capitalization, but minimum is better, particularly if the caption is a description rather than a label.

Since figures should be fairly self-contained, the caption is an appropriate place to explain important details. For example, a graph might show running time for an algorithm over various data sets; the caption could include parameter values. The caption can also be used to expand abbreviations or notation used in headings.

Examples of figures. The following pages contain example tables and figures, illustrating the principles discussed above.

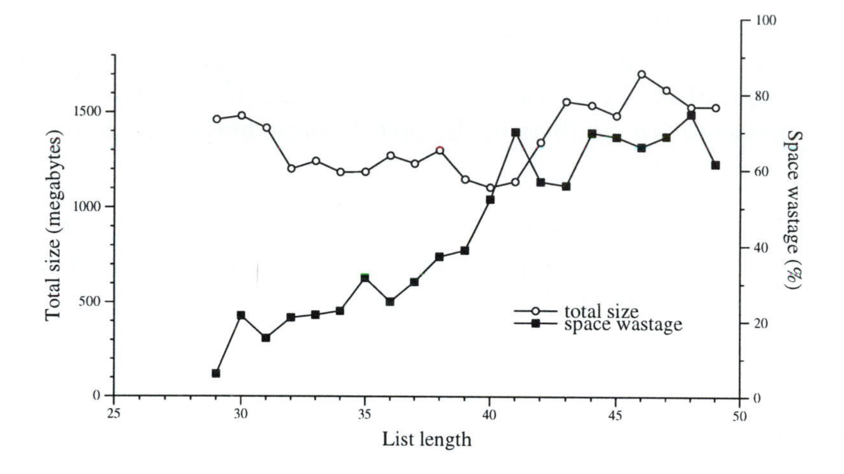

FIGURE 2. *Size and space wastage as a function of average list length.*

Two functions plotted on one graph. Note that it is necessary to label the axes to correspond with the curves; otherwise it would be difficult to identify which curve matched which axis.

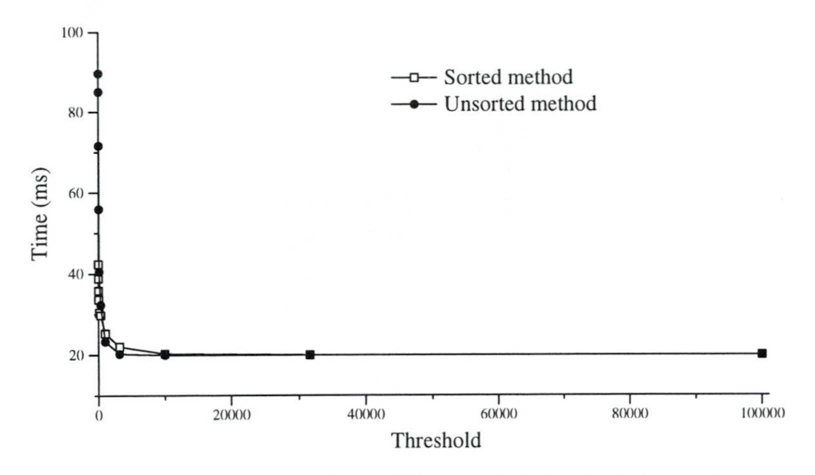

FIGURE 6. *Evaluation time (in milliseconds) for bulk insertion methods as threshold is varied.*

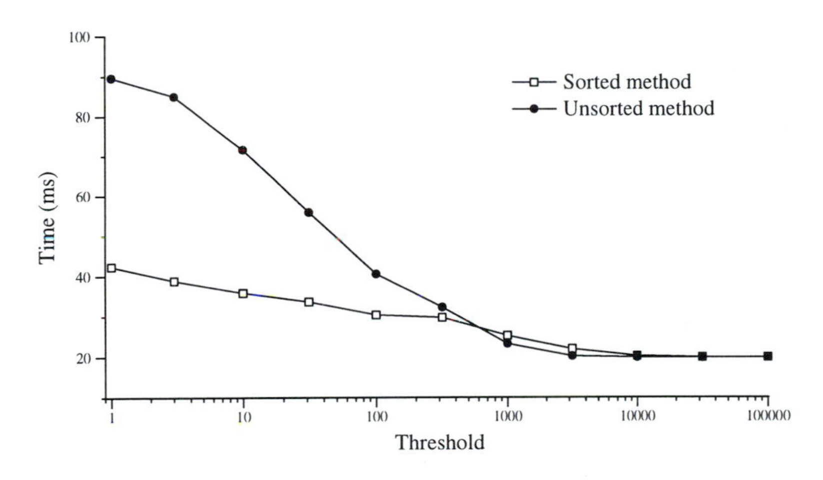

FIGURE 6. *Evaluation time (in milliseconds) for bulk insertion methods as threshold is varied.*

Choice of axis scaling. For these graphs showing the same data, in the lower graph the logarithmic scaling on the x-axis allows the behaviour for small thresholds to be seen.

Data	Method	
set	A	B
1	11.5	11.6
2	27.9	17.1
3	9.7	8.2
4	24.0	13.5
5	49.4	60.1
6	21.1	35.4
7	1.0	5.5

TABLE 2. *Processing time (milliseconds) for methods A and B applied to data sets 1–7.*

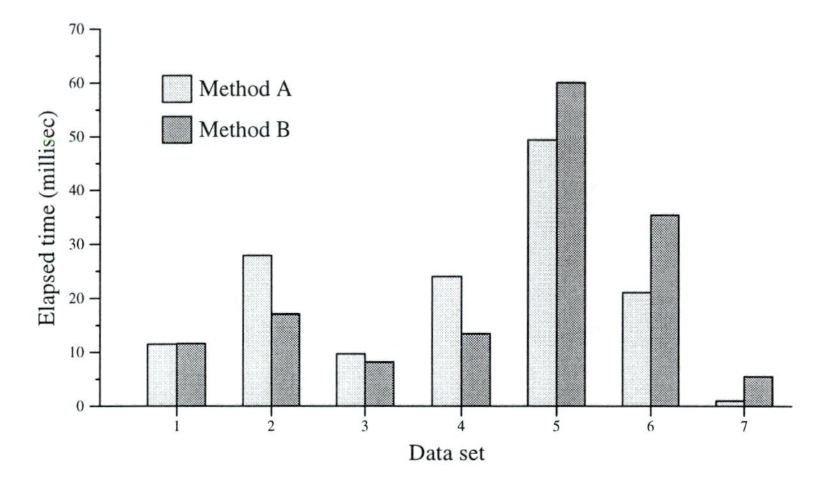

FIGURE 2. *Processing time (milliseconds) for methods A and B applied to data sets 1–7.*

Replacing a table with a graph. The data shows how two methods compare over seven experiments.

Method	Space (%)	Time (ms)
A	1.0	7 564.5
B	31.7	895.6
C	44.7	458.4
D	97.8	71.8
E	158.1	18.9
F	173.7	1.4
G	300.0	0.9

TABLE 8.4. *Tradeoff of space against time for methods A to G.*

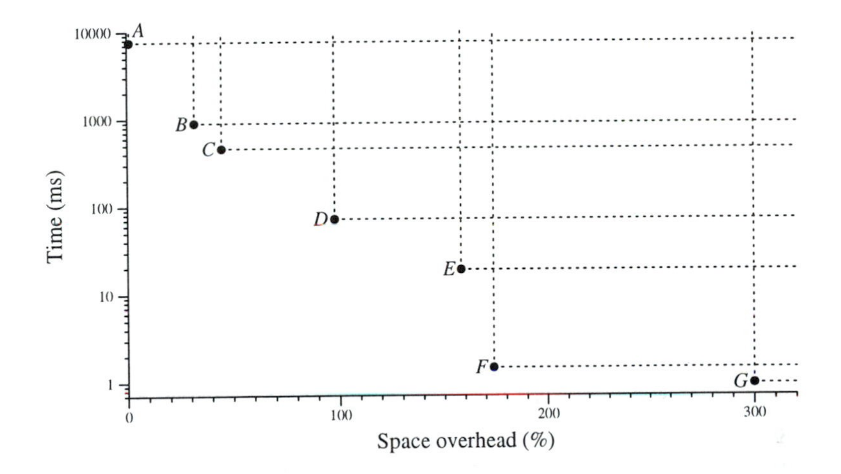

FIGURE 8.4. *Tradeoff of space against time for methods A to G. The boxed area to the right and above each point is of unacceptable performance: any method in that area will be less efficient with respect to both space and time than the point at the box's corner.*

Another table replaced by a graph. The data shows how different methods compare with respect to space and time.

STATISTICS	SMALL	LARGE
Characters	18,621	1,231,109
Words	2,060	173,145
After stopping	1,200	98,234
Index size	1.31 Kb	109.0 Kb

TABLE 6. *Statistics of text collections used in experiments.*

	Collection	
	Small	Large
File size (Kb)	18.2	1 202.3
Index size (Kb)	1.3	109.0
Number of words	2 060	173 145
After stopping	1 200	98 234

TABLE 6. *Statistics of text collections used in experiments.*

Two versions of a table. The upper table is poor. No use has been made of table hierarchy—all the elements are at the same level, so that case has to be used to differentiate between headings and content. Different units have been used for file sizes in different lines (assuming characters are one byte each). Units haven't been factored out in the last line and the precision is inconsistent. The heading of the first column is unnecessary and the table has too many horizontal lines.
In the lower table there are no vertical lines. Rows of the same type are now adjacent so that they can be compared by the reader. Note that the values of different units do not need to be vertically aligned on the decimal point or presented with the same precision.

Parameter	Data set			
	SINGLE		MULTIPLE	
	CPU (msec)	Effective (%)	CPU (msec)	Effective (%)
n ($k = 10$, $p = 100$)				
2	57.5	55.5	174.2	22.2
3	21.5	50.4	79.4	19.9
4	16.9	47.5	66.1	16.3
k ($n = 2$, $p = 100$)				
10	57.5	51.3	171.4	21.7
100	60.0	56.1	163.1	21.3
1000	111.3	55.9	228.8	21.4
p ($n = 2$, $k = 10$)				
100	3.3	5.5	6.1	1.2
1000	13.8	12.6	19.8	2.1
10 000	84.5	56.0	126.4	6.3
100 000	—	—	290.7	21.9

TABLE 2.1. *Effect on performance (processing time and effectiveness) of varying each of the three parameters in turn, for both data sets. Default parameter values are shown in parentheses. Note that $p = 100\,000$ is not meaningful for the data set* SINGLE.

Table with a deep hierarchy. There are two columns, one for parameters and one for data sets. The latter is divided into two columns, one for each data set. Each data set has two columns of figures. There are four rows, one of headings and one for each of the parameters n, k, and p. Each of these is subdivided. Note that even this rather complex table does not require vertical rules.
This table might benefit from being separated into parts, but it is helpful to have all the data together. There are insufficient data points for each parameter to justify use of a graph.

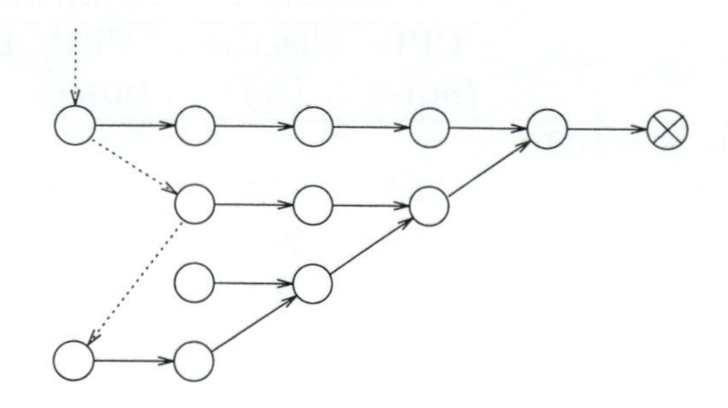

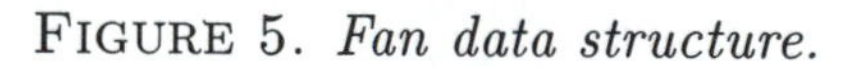

FIGURE 5. *Fan data structure.*

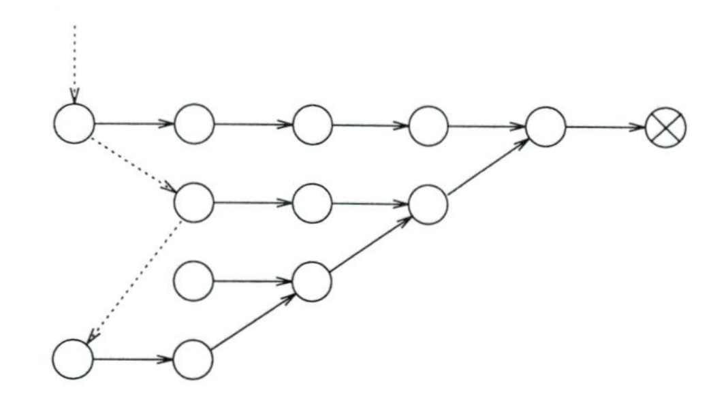

FIGURE 5. *Fan data structure, of lists with a common tail. The crossed node is a sentinel. Solid lines are within-list pointers. Dotted lines are inter-list pointers.*

Styles of caption. For these identical figures, the lower caption is preferable because it allows the figure to be less dependent on the paper's text.

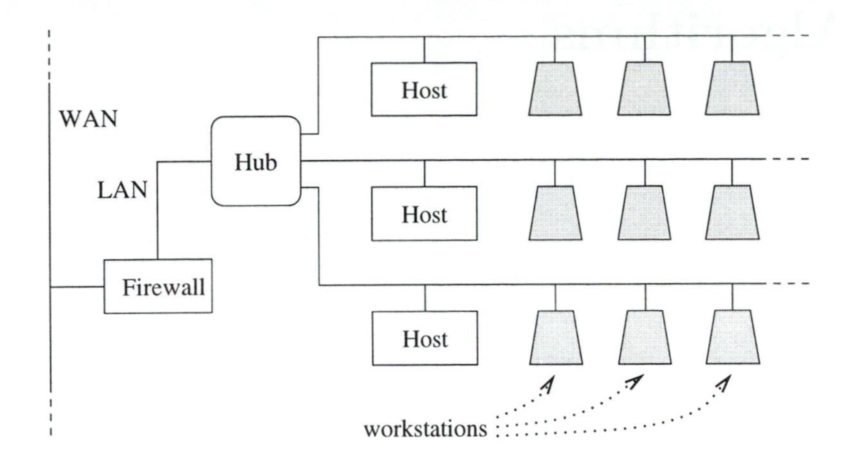

FIGURE C. *Revised network, incorporating firewall and hub with hosts and workstations on separate cables.*

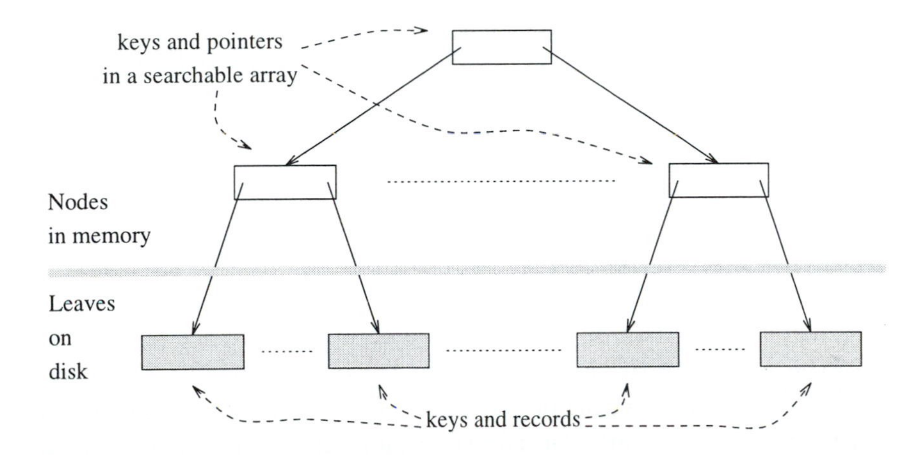

FIGURE 1.3. *Tree data structure, showing internal nodes in memory and external leaves on disk; omitted nodes are indicated by dotted lines. Nodes allow fast search and contain only keys and pointers. Leaves use compact storage and contain the records.*

Shading and dashing in diagrams. In these illustrations of structures, consistent use of shading and dashing distinguishes between different kinds of entities.

7 Algorithms

Mostly gobbledygook ...

Eric Partridge, defining computation
Usage and Abusage

Presentation of algorithms

When an algorithm is presented in a computer science paper, the details of the algorithm by themselves—the program steps, for example—do not show that it is worthwhile. The author must demonstrate that the algorithm is a worthwhile contribution: show that it is correct (given appropriate input it terminates with appropriate results) and show, by proof, experiment, or both, that it meets some claimed performance bound.

There are many reasons why an author might choose to describe an algorithm. One is that it provides a new or better way to compute a result. What is usually meant by "better" is that the algorithm can compute the result with asymptotically fewer resources as measured by a complexity analysis: less time or memory, or some desirable tradeoff of time and memory. It may be that the worst case is improved, at no saving in the average case; or that the average-case time is improved, but at the expense of space; or that all cases are improved, asymptotically, but with constant factors so large that there will no improvement in any conceivable practical situation. All of these are valid results, but it is crucial that the scope of the improvement be clearly specified—"better" is too vague.

Validation by experiment is often an important part of the presentation of algorithms. The experiment provides concrete evidence that, for some data, the algorithm terminates correctly and performs as predicted. Experiments are discussed in detail in Chapter 8.

Thus a reader would expect to find a description of some or all of:

- The steps that make up the algorithm.
- The input and output, and the internal data structures used by the algorithm.
- The scope of application of the algorithm and its limitations.
- The properties that will allow demonstration of correctness, such as preconditions, postconditions, and loop invariants.
- A demonstration of correctness.
- A complexity analysis, for both space and time requirements.
- Experiments confirming the theoretical results.

But note that while experiments on an algorithm may support an asymptotic analysis, they cannot replace it.

Another reason for describing an algorithm is to explain a complex process. For example, a paper about a distributed architecture might include a description of the steps used to communicate a packet from one processor to another. These steps certainly constitute an algorithm, and, while readers would not expect a complexity analysis, the author would have to give an argument to show that the steps did indeed result in packet transmission. Other examples are algorithms such as parsers. That is, there is no blanket requirement that a complexity analysis must be given—different norms apply to different areas and audiences—but that does not excuse authors from giving a complexity analysis where it is appropriate to do so.

Yet another reason for describing an algorithm is to show that it is feasible to compute a result, regardless of the cost, or to show that a problem is decidable. Once again different norms apply: in such cases a formal proof of correctness is essential, while an asymptotic analysis is of relatively little interest.

In summary, in the presentation of an algorithm it is usual to give a formal demonstration of correctness and performance and perhaps an experimental validation. When such demonstrations are absent, the reason for the absence should be clear.

Formalisms

The description of an algorithm usually consists of the algorithm itself and the environment required by the algorithm. There are several common formalisms for presenting algorithms. One is the list style, in which the algorithm is broken down into a series of numbered or named steps and loops involving several steps are represented by "go to step X" statements. This form has the advantage that the algorithm can be discussed as it is presented: there is no restriction on the amount of text used to describe a step (although a step should be a single activity), so there is room for a clear statement of each step and for remarks on its properties. But the control structure is often obscure and it is all too easy for the discussion to bury the algorithm.

Another common formalism is pseudocode, in which the algorithm is presented as if written in a block-structured language and each line is numbered. An example is shown on page 107. Pseudocode has the advantage that the structure of the algorithm is immediately obvious; but each statement is forced by formatting considerations to be fairly terse, and it is not easy to include detailed comments. Also, as discussed below, the use of programming language constructs and notation is usually a mistake. It takes experience to present algorithms well in pseudocode, and, although it is straightforward to translate such pseudocode into any imperative programming language, it is unnecessarily hard to understand.

A better option is to use what might be called prosecode: number each step, never break a loop over several steps, use subnumbering for the parts of a step, and include explanatory text. An example is shown on page 108. In the example, input and output are described in the preamble, and statements and explanatory text are mixed freely in the algorithm itself. Despite the informality, the specification of the algorithm is direct and clear. The assignment symbol "$\leftarrow$" is a good choice because it is unambiguous, in contrast to symbols such as "$=$". Note the use of nested labelling for nested statements. However, the prosecode style of presentation is only effective when the concepts underlying the algorithm have been discussed before the algorithm is given.

Another effective approach to description of algorithms is what might be called literate code, in which the detail of the algorithm is introduced gradually, intermingled with discussion of the underlying ideas and perhaps with the asymptotic analysis and proof of correctness. An example is shown on page 109. (This example is incomplete—most algorithms worth presenting need a substantial explanation that can't necessarily be condensed into a page or two.)

Flowcharts should not be used to describe algorithms, for many reasons: lack of modularity, promotion of the use of `goto` statements, lack of space for explanatory text, insufficient space for complex conditions, and inability to clearly represent algorithms of any complexity.

Notation

Mathematical notation is preferable to programming notation for presentation of algorithms. While the expression "`x[i]`" might be used in a program, in a technical article it is almost always better to use "x_i". Don't use "`*`" or "x" to denote multiplication; most word processors provide a multiplication symbol such as "$\times$" or "$\cdot$", and in any case multiplication is often implicit. Likewise, avoid using constructs from specific programming languages. For example, expressions such as `==`, `a = b = 0`, `a++`, and `for ( i=0 ; i<n ; i++ )` may have little meaning, or even the wrong meaning, to readers who are unfamiliar with C. Block-bounding statements such as `begin` and `end` are usually unnecessary; nesting can be shown by indentation or by the numbering style, as in the examples on pages 107 and 108.

Mathematics provides many handy conventions and symbols that can be used in description of algorithms, including set notation, subscripts and superscripts, symbols such as $\lceil$ and $\rceil$, $\sum$, $\prod$, and so on. But remember that such notation has a widely understood formal meaning that should not be abused. Also, good programming style does not necessarily imply good style for description of algorithms. For example, take care with variable names of more than one character—don't use "pq" if it might be interpreted as "$p \times q$".

It was once common to include the text of a program in an article, in addition to a description of the algorithm it embodies. This was valuable because, for short programs at least, it was the simplest way for readers to obtain the code. However, there are now better ways of making code available (such as ftp or http) and few readers are eager to key in a program of any size.

Level of detail

Algorithms should be specified in sufficient detail to allow them to be implemented without undue inventiveness.

× 5. (Matching.) For each pair of strings $s, t \in S$, find $N_{s,t}$, the maximum number of non-overlapping substrings that s and t have in common.

The way in which a step of this kind is implemented may greatly affect the behaviour of the final algorithm, so the matching process needs to be made explicit. But don't provide too much detail. For example, loops are sometimes used unnecessarily in specification of algorithms.

× 3. (Summation.) Set $\mathit{sum} \leftarrow 0$. For each j, where $1 \leq j \leq n$,
   (a) Set $c \leftarrow 1$; the variable c is a temporary accumulator.
   (b) For each k, where $1 \leq k \leq m$, set $c \leftarrow c \times A_{jk}$.
   (c) Set $\mathit{sum} \leftarrow \mathit{sum} + c$.

This is poor because it is cumbersome and no more informative than the equivalent mathematical expression. It is safe to assume that most programmers know how to use loops to implement sums and products.

✓ 3. (Summation.) Set $\mathit{sum} \leftarrow \sum_{j=1}^{n} \left(\prod_{k=1}^{n} A_{jk} \right)$.

As this form of the step illustrates, the step is probably unnecessary unless *sum* is used more than once: use of *sum* could be replaced by the summation it represents. The one reason to have a step just for the summation would be to include explanation of any difficult issues; for example, if the matrix A was sparse and stored as a list rather than a two-dimensional array, there might be an explanation of how to compute the summation efficiently.

In specifications of algorithms, use English rather than mathematics if the former is sufficiently clear.

× 2. for $1 \leq i \leq |s|$
   (a) set $c \leftarrow s[i]$
   (b) set $A_c \leftarrow A_c + 1$

✓ 2. For each character c in string s, increment A_c.

Figures

Figures are an effective way of conveying the intricacies of data structures; and even quite simple structures can require complex descriptions. General guidelines for figures are given in Chapter 6.

✓ A single rotation can be used to bring a node one level closer to the root. In a left-rotation, a node x and its right child y are exchanged as follows: given that B is the left child of y, then assign B to be the new right child of x and assign x to be the new left child of y. The reverse operation is a right-rotation. Left- and right-rotations are shown in the following diagram.

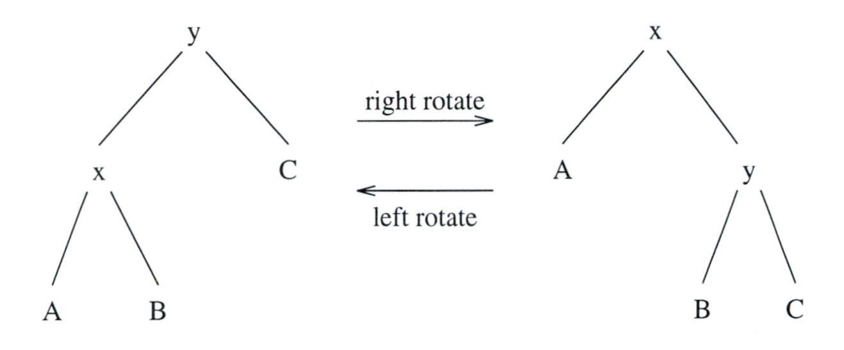

Environment of algorithms

The steps that comprise an algorithm are only part of its description. The other part is its environment: the data structures on which it operates, input and output data types, and, in some cases, factors such as properties of the underlying operating system and hardware. If the environment of an algorithm is not clearly described the algorithm is likely to be difficult to understand. For example, a presentation of a list-processing algorithm should include descriptions of the list type, the input, and the possible outputs. If the list is stored on secondary storage and speed is being analyzed, it might also be appropriate to describe assumed disk characteristics. For algorithms in which there are hardware considerations, such as memory size or disk throughput, for the environment to seem realistic any assumptions about the hardware should reflect current technology or likely improvements in the near future.

Specify the types of all variables, other than trivial items such as counters; describe expected input and output, including assumptions about the correctness of the input; state any limitations of the algorithm; and discuss possible errors that are not explicitly captured by the algorithm. Most importantly, say what the algorithm does.

Describe data structures carefully. I do not mean that you should give record definitions in a pseudo-language; instead, use, say, a simple mathematical notation to unambiguously specify the structure.

✓ Each element is a triple

```
( string, length, positions )
```

in which `positions` is a set of byte offsets at which `string` has been observed.

Be consistent. When presenting several algorithms for the same task, they should as far as possible be defined over the same input and output. It may be the case that some of the algorithms will be more powerful than the others—can process a richer input language, for example. Variations of this kind should be made explicit.

Performance of algorithms

The tools for evaluating the performance of algorithms, and for comparing algorithms, are formal proof, mathematical modelling, simulation, and experimentation. These and other issues related to testing are discussed in Chapter 8. Here I discuss the aspects of algorithms that might be considered in an evaluation.

Basis of evaluation. The basis of evaluation should be made explicit. Where algorithms are being compared, specify not only the environment but also the criteria used for comparison. For example, are the algorithms being compared for functionality or speed? Is speed to be examined asymptotically or for typical data? Is the data real or synthetic? When describing the performance of a new technique, it is helpful to compare it to a well-known standard.

Readers will be more convinced by a comparison that has a realistic basis. In particular, the basis should not appear to favour the algorithm being demonstrated over existing algorithms—if the basis of comparison is questionable, so too will the results be questionable.

Simplifying assumptions can be used to make mathematical analysis tractable, but can give unrealistic models. Non-trivial simplifications should be carefully justified.

Processing time. Time (or speed) over some given input is one of the principal resources used by algorithms; the others are memory, disk space, disk traffic, and network traffic. Time is not always an easy quantity to measure, since it depends on factors such as CPU speed, cache sizes, system load, and hardware dependencies such as prefetch strategy. Nonetheless, some absolute indication of time should be part of the description of any new algorithm. Times based on a mathematical model rather than on experiment should be clearly indicated as such.

Measurements of CPU time can be unreliable. CPU times in most systems are counted as multiples of some fixed fraction of a second, say a sixty-fourth or a thousandth. Each of these fractions of time is allocated to a process, often by heuristics such as simply choosing the process that is active at that moment. Thus the reported CPU time for a process may be no more than a good estimate, particularly if the system is busy.

Memory and disk requirements. It is often possible to trade memory requirements against time, not only by choice of algorithm but also by changing the way disk is used. Authors should take care to specify how their algorithms use memory.

Disk and network traffic. Disk costs have two components, the time to fetch the first bit of requested data (seek and latency time) and the time required to transmit the requested data (at a transfer rate). Thus sequential accesses and random accesses have greatly different costs. For current hardware, in which there are several levels of caching and buffering between disk and user process, it may also be appropriate to consider repeat accesses, in which case there is some likelihood that the access cost will be low. The behaviour of network traffic is similar—the cost of transmitting the first byte is greater than the cost for subsequent bytes, for example.

Because of the sophistication of current disk drives and the complexity of their interaction with CPU and operating system, exact mathematical descriptions of algorithm behaviour are unattainable; broad approximations are often the only manageable way of describing disk performance.

Applicability. Algorithms can be compared not only with regard to their resource requirements, but with regard to functionality. The basis of such comparisons will be quite different to those based on, say, asymptotic analysis.

A common error is to compare the resource requirements of two algorithms that perform subtly different tasks. For example, the various approximate string matching algorithms do not yield the same results—strings that are alike according to one algorithm can be unalike according to another. Comparing the costs of these algorithms is not particularly informative.

Asymptotic complexity

The performance of algorithms is often measured by asymptotic analysis; the reader should learn how an algorithm behaves as the scale of the problem changes. Most readers are familiar with big-O notation: a function $f(n)$ is said to be $\mathrm{O}(g(n))$ if for some constants c and k we have $f(n) \leq c \cdot g(n)$ for all $n > k$, or is said to be $\mathrm{o}(g(n))$ if $f(n) < c \cdot g(n)$ for all $n > k$. Thus big-O notation is used to describe an upper bound. Likewise, Ω and ω are used to describe lower bounds, and Θ is used to describe tight bounds—a function has complexity $\Theta(g(n))$ if it has complexities $\Omega(g(n))$ and $\mathrm{O}(g(n))$. Some authors use this notation in slightly different ways, so it is helpful to define what you mean by, for example, $\Omega(g(n))$.

Big-O notation is also used in another, less formal sense, to mean *the complexity* rather than *an upper bound on the complexity*. An author might write that "comparison-based sorting takes $O(n \log n)$ time" or that "linear insertion sort always takes at least $O(n)$ time"; which, although an abuse, is perfectly clear and has stronger emphasis than "linear insertion sort has complexity $\Omega(n)$". But beware of loose usage that could be misunderstood. When an author describes an algorithm as "quadratic", some readers may assume that complexity $\Theta(n^2)$ is meant, while others make a different interpretation. Similarly be careful with "constant", "linear", "logarithmic", and "exponential".

For algorithms that operate on static data structures, it may be appropriate to consider the cost of creating that data structure. For example, binary search in a sorted array takes $O(\log n)$ time, but $O(n \log n)$ time is required to initially sort the array.

Make sure that the domain of the analysis is clear, and be careful to analyze the right component of the data. It would usually be appropriate, for example, to analyze database algorithms as a function of the number of records, not of the length of individual records. However, if record length can substantially vary then it too should be considered. For algorithms that apply arithmetic to integers it may be appropriate to regard each arithmetic operation as having unit cost. On the other hand, if the integers involved can be of arbitrary length—consider for example public-key encryption algorithms that rely for privacy on the expense of prime factorization—it is appropriate to regard the cost of the arithmetic operations as a function of the number of bits in each integer.

Subtle problems are that the dominant cost may change with scale, and that the cost that is dominant in theory may never dominate in practice. For example, a certain algorithm might require $O(n \log n)$ comparisons and $O(n)$ disk accesses. In principle the complexity of the algorithm is $O(n \log n)$, but, given that a disk access may require 10 milliseconds and a comparison 50 nanoseconds, in practice the cost of the disk accesses might well dominate for any possible application.

Some authors misunderstand the logic of asymptotic claims. For example, Amdahl's law states that the lower bound for the time taken for an algorithm to complete is determined by the part of the algorithm that is inherently sequential. The remainder can be executed in parallel and hence time for this part can be reduced by addition of processors, but no increase in the number of processors can affect the lower bound. However, in a published paper it was claimed that Amdahl's law was broken by, for a certain algorithm, increasing both the size of the input data and the number of processors. These changes had minimal impact on the sequential part of the algorithm, so that the proportion of total processing time spent in the sequential part was reduced; but this result does not contradict Amdahl's law, and so the claim was false.

Another fallacious claim was that, for a certain indexing technique, the time required to find matches to a pattern in a database was asymptotically sublinear in the database size—a remarkable result, because the probability that a record is a match to a given pattern is fixed, so that in the limit the number of matches must be linear in database size. The error was that the author had assumed that the length of the pattern was a logarithmic function of database size, so that the number of answers was constant. The technique gave the appearance of being sublinear because the input was changing.

Sometimes a formal analysis is inappropriate or only a minor consideration. For example, an algorithm for introducing line breaks into paragraphs of text will only rarely have to operate on a large input, so showing that a new algorithm is better than an existing algorithm in the limit may be of less interest than showing it is better on a typical case. More generally, although some results can be conclusively obtained by analysis, others cannot. Analytical results often say nothing about constant factors, for example, or behaviour in practice where CPU, cache, bus, and disk can interact in unpredictable ways. Such properties can only be determined by experiment. Thus, while an asymptotic analysis tells us that a hash table should be faster than a B-tree, in practice the B-tree may be superior for storage of records in a large database system.

Moreover, an analysis is no more reliable than its assumptions. In an analysis of a data structure, the data must be modelled in some way, perhaps with simplifying assumptions to make the analysis tractable; but there is no guarantee that the modelling is realistic. Analytic results can be powerful indeed—with, in some cases, implications for performance in practice on all machines for all time—but they are not necessarily sufficient by themselves.

Example of pseudocode. This is not the best style of presentation: the algorithm is cryptic and the numbering does not reflect the indentation. Also, the author has unnecessarily introduced a trivial optimization (at lines 10 and 12) and the notation for variables is ugly. It is like a program meant for a machine, not an explanation meant for a reader.

The **WeightedEdit** function computes the edit distance between two strings, assigning a higher penalty for errors closer to the front.

Input:	$S1, S2$: strings to be compared.
Output:	weighted edit distance
Variables:	$L1, L2$: string lengths
	$F[L1, L2]$: array of minimum distances
	W: current weighting
	M: maximum penalty
	C: current penalty

WeightedEdit$(S1, S2)$:

1. $L1 = len(S1)$
2. $L2 = len(S2)$
3. $M = 2 \times (L1 + L2)$
4. $F[0, 0] = 0$
5. **for** i **from** 1 **to** $L1$
6. $\quad F[i, 0] = F[i-1, 0] + M - i$
7. **for** j **from** 1 **to** $L2$
8. $\quad F[0, j] = F[0, j-1] + M - j$
9. **for** i **from** 1 **to** $L1$
10. $\quad C = M - i$
11. $\quad$ **for** j **from** 1 **to** $L2$
12. $\qquad C = C - 1$
13. $\qquad F[i, j] = min(F[i-1, j] + C,$ $F[i, j-1] + C,$ $F[i-1, j-1] + C \times isdiff(S1[i], S2[j]))$
14. **WeightedEdit** $= F[L1, L2]$

Example of prosecode. The longer introduction and use of text in the presentation help make the algorithm easy to understand.

WeightedEdit(s, t) compares two strings s and t, of lengths k_s and k_t respectively, to determine their edit distance—the minimum cost in insertions, deletions, and replacements required to convert one into the other. These costs are weighted so that errors near the start of the strings attract a higher penalty than errors near the end.
We denote the ith character of string s by s_i. The principal internal data structure is a 2-dimensional array F in which the dimensions have ranges 0 to k_s and 0 to k_t respectively. When the array is filled, $F_{i,j}$ is the minimum edit distance between the strings $s_1 \ldots s_i$ and $t_1 \ldots t_j$; and F_{k_s,k_t} is the minimum edit distance between s and t.
The value p is the maximum penalty, and the penalty for a discrepancy between positions i and j of s and t respectively is $p - i - j$, so that the minimum penalty is $p - k_s - k_t = p/2$ and the next-smallest penalty is $p/2 + 1$. Two errors, wherever they occur, will outweigh one.

1. (Set penalty.) Set $p \leftarrow 2 \times (k_s + k_t)$.
2. (Initialize data structure.) The boundaries of array F are initialized with the penalty for deletions at start of string; for example, $F_{i,0}$ is the penalty for deleting i characters from the start of s.
   (a) Set $F_{0,0} \leftarrow 0$.
   (b) For each position i in s, set $F_{i,0} \leftarrow F_{i-1,0} + p - i$.
   (c) For each position j in t, set $F_{0,j} \leftarrow F_{0,j-1} + p - j$.
3. (Compute edit distance.) For each position i in s and position j in t,
   (a) The penalty is $C = p - i - j$.
   (b) The cost of inserting a character into t (equivalently, deleting from s) is $I = F_{i-1,j} + C$.
   (c) The cost of deleting a character from t is $D = F_{i,j-1} + C$.
   (d) If s_i is identical to t_j, the replacement cost is $R = F_{i-1,j-1}$. Otherwise, the replacement cost is $R = F_{i-1,j-1} + C$.
   (e) Set $F_{i,j} \leftarrow \min(I, D, R)$.
4. (Return.) Return F_{k_s,k_t}.

Example of literate code. The algorithm is explained and presented simultaneously. This is the most verbose style, but, usually, the clearest. Note that this example is incomplete.

WeightedEdit(s, t) compares two strings s and t, of lengths k_s and k_t respectively, to determine their edit distance—the minimum cost in insertions, deletions, and replacements required to convert one into the other. These costs are weighted so that errors near the start of the strings attract a higher penalty than errors near the end.

The major steps of the algorithm are as follows.

1. Set the penalty.
2. Initialize the data structure.
3. Compute the edit distance.

We now examine these steps in detail.

1. Set the penalty.

   The main property that we require of the penalty scheme is that costs reduce smoothly from start to end of string. As we will see, the algorithm proceeds by comparing each position i in s to each position j in t. Thus a diminishing penalty can be computed with $p - i - j$, where p is the maximum penalty. By setting the penalty with

   (a) Set $p \leftarrow 2 \times (k_s + k_t)$

   the minimum penalty is $p - k_s - k_t = p/2$ and the next-smallest penalty is $p/2 + 1$. This means that two errors—regardless of position in the strings—will outweigh one.

2. Initialize data structures ...

8 Hypotheses and experiments

Even the clearest and most perfect circumstantial evidence is likely to be at fault, after all, and therefore ought to be received with great caution.

Mark Twain
Pudd'nhead Wilson's Calendar

There are as many scientific methods as there are individual scientists.

Percy W. Bridgman
On "Scientific Method"

We never know what we are talking about.

Karl Popper
Unended Quest

The use of experiments to verify hypotheses is one of the central elements of science. A hypothesis is a

> ... trial idea, a tentative suggestion concerning the nature of things. Until it has been *tested*, it should not be confused with a *law* ... Plausibility is not a substitute for evidence, however great may be the emotional wish to believe. [25]

In computing, experiments—most commonly an implementation tried against test data—are used for purposes such as verifying hypotheses about algorithms: not just showing that they perform the specified task, but can do so with appropriate resources. A tested hypothesis will become part of scientific knowledge if it is sufficiently well described and constructed, and if it is convincingly demonstrated. Here I consider how to frame hypotheses and how to design and describe experiments, but be aware that there is no neat formula for experimentation, just as there is no neat formula for "the process of research".

Stating hypotheses

A hypothesis should be specified clearly and precisely, and should be unambiguous. The more loosely a concept is defined, the more easily it will satisfy many needs simultaneously, even when these are contradictory. Often it is important to state what is *not* being proposed—what the limits on the conclusions will be. Consider an example. Suppose P-lists are a well-known data structure used for a range of applications, in particular as an in-memory search structure that is fast and compact. A scientist has developed a new data structure called the Q-list. Formal analysis has shown the two structures to have the same asymptotic complexity in both space and time but the scientist intuitively believes the Q-list to be superior in practice, and has decided to demonstrate this by experiment.

(This motivation by belief, or instinct, is a crucial element of the process of science: since ideas cannot be known to be correct when first conceived, it is intuition or plausibility that suggests them as worthy of consideration. That is, the investigation may well have been undertaken for subjective reasons; but the final report on the research, the published paper, must be objective.)

The hypothesis might be encapsulated as

× Q-lists are superior to P-lists.

But this statement does not suffice as the basis of experiment: success would have to apply in all applications, in all conditions, for all time. Formal analysis might be able to justify such a result, but no experiment will be so far-reaching. In any case, it is rare indeed for a data structure to be completely superseded—consider the durability of arrays and linked lists—so in all probability this hypothesis is incorrect. A testable hypothesis might be

✓ As an in-memory search structure for large data sets, Q-lists are faster and more compact than P-lists.

Further qualification may well be necessary.

✓ We assume there is a skew access pattern, that is, that the majority of accesses will be to a small proportion of the data.

The qualifying statement imposes a scope on the claims made on behalf of Q-lists. A reader of the hypothesis has enough information to reasonably conclude that Q-lists do not suit a certain application, which in no way invalidates the result. Another scientist would be free to explore the behaviour of Q-lists under another set of conditions, in which they might be inferior to P-lists, but again the original result remains valid.

As the example illustrates, a hypothesis must be testable. One aspect of testability is that the scope be limited to a domain that can feasibly be explored. Another, crucial aspect is that the hypothesis should be capable of falsification. Vague claims are unlikely to meet this criterion.

✗ Q-list performance is comparable to P-list performance.

✗ Our proposed query language is relatively easy to learn.

The more vulnerable a hypothesis is to falsification, the more convincing is a successful demonstration.

Developing hypotheses

It may be necessary to refine a hypothesis as a result of initial testing; indeed, much of scientific progress can be viewed as refinement and development of hypotheses to fit new observations. Occasionally there is no room for refinement, a classic example being Einstein's prediction of the deflection of light by massive bodies—a hypothesis much exposed to disproof, since it was believed that significant deviation from the predicted value would invalidate the theory of general relativity. But more typically a hypothesis will evolve in tandem with refinements in the experiments.

This is not, however, to say that the hypothesis should follow the experiments. A hypothesis will often be based on observations, but can only be regarded as confirmed if able to make successful predictions. There is a vast difference between an observation such as "the algorithm worked on our data" and a tested hypothesis such as "the algorithm was predicted to work on any data of this class, and this prediction has been confirmed on our data". Another way of regarding this issue is that, as far as possible, tests should be blind. If an experiment and hypothesis have been fine-tuned on the data, it cannot be said that the experiment provides confirmation. At best the experiment has provided observations on which the hypothesis is based.

Where two hypotheses fit the observations equally well and one is clearly simpler than the other, the simpler should be chosen. This principle, known as Occam's razor, is purely a convention; but it is well-established and there is certainly no reason to choose a complex explanation when another is at hand.

Defending hypotheses

One component of a strong paper is a precise, interesting hypothesis. Another component is the testing of the hypothesis and the presentation of the supporting evidence. As part of the research process you need to test your hypothesis and if it is correct (or, at least, not falsified) assemble supporting evidence. For the presentation of the hypothesis you need to construct an argument relating your hypothesis to the evidence.

For example, the hypothesis "the new range searching method is faster than previous methods" might be supported by the evidence "range search amongst n elements requires $O(\log \log n)$ comparisons". This may or may not be good evidence, but it is not convincing because there is no argument connecting the evidence to the hypothesis. What is missing is information such as "previous results indicated a theoretical best-case complexity of $O(\log n)$". It is the role of the connecting argument to show that the evidence does indeed support the hypothesis, and to show that conclusions have been drawn correctly.

In constructing an argument is can be helpful to imagine yourself defending your hypothesis to a colleague, so that you play the role of inquisitor. That is, raising objections and defending yourself against them is a way of gathering the material needed to convince the reader that your argument is correct. Starting from the hypothesis that "the new string hashing algorithm is fast because it doesn't use multiplication or division" you might debate as follows:

– I don't see why multiplication and division are a problem.

  On most machines they use several cycles, or may not be implemented in hardware at all. The new algorithm instead uses two exclusive-or operations per character and a modulo in the final step. I agree that for pipelined machines with floating-point accelerators the difference may not be great.

– Modulo isn't always in hardware either.

  True, but it is only required once.

– So there is also an array lookup? That can be slow.

  Not if the array is cache-resident.

– What happens if the hash table size is not 2^8?

  Good point. This function is most effective for hash tables of size 2^8, 2^{16}, and so on.

In your argument you need to rebut likely objections while conceding points that can't be rebutted and admitting when you are uncertain. If, in the process of developing your hypothesis, you raised an objection but reasoned it away, it can be valuable to include the reasoning in the paper. Doing so helps the reader to follow your train of thought, and certainly helps the reader who independently raises the same objection. That is, you need to anticipate problems the reader may have with your hypothesis. Likewise, you should actively search for counter-examples.

If you think of an objection that you cannot refute, don't just put it aside. At the very least you should raise it yourself in the paper, but it may well mean that you must reconsider your results.

A hypothesis can be tested in a preliminary way by considering its effect, that is, by examining whether there is a simple argument for keeping or discarding it. For example, are there any improbable consequences if the hypothesis is true? If so, there is a good chance that the hypothesis is wrong. For a hypothesis that displaces or contradicts some currently-held belief, is the contradiction such that the belief can only have been held out of stupidity? Again, the hypothesis is probably wrong. Does the hypothesis cover all of the observations explained by the current belief? If not, the hypothesis is probably uninteresting.

Always consider the possibility that your hypothesis is wrong. It is often the case that a correct hypothesis at times seems dubious—perhaps initially, before it is fully developed, or when it appears to be contradicted by some experimental evidence—but the hypothesis survives and is even strengthened by test and refinement in the face of doubt. But equally often a hypothesis is false, in which case clinging to it is a waste of time. Persist for long enough to establish whether or not it is likely to be true, but to persist longer is foolish.

A corollary is that the stronger your intuitive liking for a hypothesis, the more rigorously you should test it—attempt to disprove it—rather than twist results, and yourself, defending it.

Evidence

There are broadly speaking four kinds of evidence that can be used to support a hypothesis: analysis or proof, modelling, simulation, and experiment.

An analysis or proof is a formal argument that the hypothesis is correct. It is a mistake to suppose that the correctness of a proof is absolute—confidence in a proof may be high but that does not guarantee that it is free from error. (In my experience it is not uncommon for a researcher to feel certain that a theorem is correct but have doubts about the mechanics of the proof, which all too often leads to the discovery that the theorem is wrong after all.) It is a mistake to suppose that all hypotheses are amenable to formal analysis, particularly hypotheses that involve the real world in some way. For example, human behaviour is intrinsic to questions about interface design, and system properties can be intractably complex. It is also a mistake to suppose that a complexity analysis is always sufficient. Nonetheless, the possibility of formal analysis should never be overlooked.

A model is a mathematical description of the hypothesis—or some component of the hypothesis such as an algorithm whose properties are being considered—and there will usually be a demonstration that the hypothesis and model do indeed correspond.

A simulation is usually an implementation or partial implementation of a simplified form of the hypothesis, in which the difficulties of a full implementation are sidestepped by omission or approximation. At one extreme a simulation might be skeletal, so that for example a parallel algorithm could be tested on a sequential machine by use of an interpreter that counts machine cycles and communication costs between simulated processors; at the other extreme a simulation could be an implementation of the hypothesis, but tested by artificial data. I think of a simulation as being a "white coats" test: artificial, isolated, conducted in a tightly-controlled environment.

An experiment is a full test of the hypothesis, based on an implementation of the proposal and on real—or at least realistic—data. Thus in an experiment there is a sense of *really doing it* while in a simulation there is a sense of *only pretending*. Ideally an experiment should be conducted in the light of predictions made by a model, so that it confirms some expected behaviour.

These methods can be used to confirm each other, with say a simulation used to provide further evidence that a proof is correct. But they should not be confused with each other. For example, suppose that for some algorithm there is a mathematical model of expected performance. Encoding this model in a program and computing predicted performance for certain values of the model parameters is in no way an experimental test of the algorithm; it does not even confirm that the model is a description of the algorithm. At best it confirms claimed properties of the model. The distinction between simulation and experiment is however more blurry.

When choosing which of these methods to use you need to consider how convincing they will be to the reader. If your evidence is questionable—say a model based on simplifications and assumptions, an involved algebraic analysis and application of advanced statistics, or an experiment on limited data—the reader may well be skeptical of the result. Select a form of evidence, not so as to keep your own effort to a minimum, but to be as persuasive as possible.

Designing fair experiments

Tests should be fair rather than constructed to support the hypothesis. This problem is arguably most acute when a new idea is being compared to an existing one. In this case, the test environment should be designed to be seen as reasonable by readers who support the existing idea—if the tests seem biased towards the new idea, these readers will not be persuaded by the results.

When considering what experiments to try, you should identify the cases in which the hypothesis is least likely to hold. These are the interesting cases: if they are not tested—if only the cases where the hypothesis is most likely to hold are tested—then the experiments won't prove much at all. The experiment should be of course be a test of the hypothesis; you should always verify that what you are testing is what you intended to be test. In particular, an experiment should only succeed if the hypothesis is correct.

When checking experimental design or outcomes, it is worth considering whether there are other possible interpretations of the results; and if so, designing further tests to eliminate these possibilities. Consider for instance the problem of finding whether a file stored on disk contains a given string. One algorithm directly scans the file; another algorithm, which has been found to give faster response, scans a compressed form of the file. Further tests would be needed to identify whether the speed gain was because the second algorithm used fewer machine cycles or because the compressed file was fetched more quickly from disk.

Care is particularly needed when checking the outcome of negative or failed experiments. A reader of the statement "we have shown that it is not possible to make further improvement" may wonder whether what has really been shown is that the author is not competent to make further improvement. That is, design of experiments to demonstrate failure is particularly difficult.

It is also worth considering whether the results obtained are sensible. For example, are there rules of conservation that should apply to this experiment? Sometimes boundary conditions are highly predictable—do the results appear to be right as they approach the boundaries? For a typical case it should be possible to make a rough guess as to expected outcome—is this observed?

Conclusions should be sufficiently supported by the results. Success in a special case does not prove success in general, so be aware of factors in the test that may make it special. A common problem is scale—whether the same result would be observed with a larger data set, for example.

Don't draw undue conclusions. If, say, one method is faster than another on a large data set, and they are of the same speed on a medium data set, that does not imply that the second is faster on a small data set—it only implies that different costs dominate at different scales. Also, don't overstate the conclusions. For example, if a new algorithm is somewhat worse than an existing one, it is wrong to describe them as similar. A reader might infer that they are similar if the difference is slight, but it is not honest for the author to make that claim.

Designing robust experiments

The test environment should be designed to minimize the effect of extraneous factors.[10] Even elementary properties can be surprisingly hard to measure: for example, access time to material stored on disk is not just a property of disk hardware, but is affected by access pattern, presence and size of disk cache, file system design, and so on. Tests should be designed to yield results that are independent of properties such as system characteristics or constant-factor overheads that are not part of the hypothesis.

For example, consider the measurement of performance of two compression techniques. If tested on different data, the results will be incomparable: we have no way of knowing whether the better performance is due to use of a better method, or due to choice of data that is inherently more compressible. Thus one particular component of a test environment is choice of test data. For some experiments standard data is available, such as benchmark problems in machine learning or the corpora used to test compression methods. The use of such standard resources is essential to experimentation on these problems. Where standard data is not available, care should be taken to ensure that the chosen test data is representative.

---

[10]In careful research published in 1648, Jan-Baptista van Helmont concluded that plants consist of water:

> That all plants immediately and substantially stem from the element water alone I have learnt from the following experiment. I took an earthen vessel in which I placed two hundred pounds of earth dried in an oven, and watered with rain water. I planted in it the stem of a willow tree weighing five pounds. Five years later it had developed into a tree weighing one hundred and sixty-nine pounds and about three ounces. Nothing but rain (and distilled water) had been added. The large vessel was placed in earth and covered by an iron lid with a tin-surface that was pierced with many holes [to allow the soil to breathe while preventing dust from adding to it –JZ]. I have not weighed the leaves that came off in the four autumn seasons. Finally I dried the earth in the vessel again and found the same two hundred pounds of it diminished by about two ounces. Hence one hundred and sixty-four pounds of wood, bark and roots had come up from water alone.

Another component of many test environments is the hardware. It is best to describe performance in terms of the characteristics of some commonly available hardware, as for example specified by clock speed, disk access time, and so on. This allows readers to relate published results to observed performance on another system.

The tests themselves should as far as possible not depend on accuracy of measurements or quality of the implementation. Ideally an experiment should be designed to yield a result that is unambiguously either true or false; where this is not possible, another form of confirmation is to demonstrate a trend or pattern of behaviour. That is, success or otherwise should be obvious, not subject to interpretation. An example of this principle is the work of Pons and Fleischman on cold fusion. Their claims of success were founded on small discrepancies—only a few percent—between measured input energy and output energy. Admission of only a small experimental error would confound their claims. In contrast, the claims that they had failed were based on the almost complete absence of a particular form of radiation, effectively a straightforward Boolean test whose result was false.

Another example of this principle is provided by the various improvements that can be made to the standard quicksort algorithm, such as better choice of pivot and loops that avoid expensive procedure calls. With test data chosen to exercise the various cases—such as initially unsorted, initially sorted, data sets with many repetitions of some values—experiments can show that the improvements do indeed lead to faster sorting. What such experiments cannot show is that quicksort is inherently better than, say, mergesort. While it might, for example, be possible to deduce that the same kinds of improvement do not yield benefits for mergesort, nothing can be deduced about the relative merits of the algorithms because the relative quality of the implementations is unknown, and because the data has not been selected to examine trends such as asymptotic performance.

In some circumstances it is possible for an experiment to succeed, or at least appear to succeed, by luck; there might be an atypical pattern to the data, or variations in system response might favour one run over another. Where such variations are possible, many runs should be made, to reduce the probability of chance success and (in the statistical sense) to give confidence in the results. This is particularly true for timings, which can be affected by other users, system overheads, inability of most operating systems to accurately allocate clock cycles to processes, and caching effects. For example, consider the apparently simple experiment of measuring how fast a block can be accessed from a file stored on disk. Under a typical operating system, the first access is slow, because locating the first block of a file requires that header blocks be fetched first; but subsequent accesses to the same block is extremely fast, because in all likelihood it will be cached in memory. Some deviousness will be required to ensure that averages over a series of runs are realistic. Now consider a more typical experiment: real time taken to evaluate queries to a database system. If the queries are poorly chosen, the times will vary because of the block caching, and multiple runs will not give realistic figures.

For speed experiments based on a series of runs, the published figures will need to be either minimum, average, median, or maximum times. Maximum times can include anomalies, for example a run during which a greedy process such as a tape dump shuts out other processes. Minimums can be underestimates, for example when the time slice allocated to a process does not include any clock ticks. But nor are averages always appropriate—outlying points may be the result of system dependencies.

Published results may include some anomalies or peculiarities. These should be explained or at least discussed. Don't discard them unless you are quite certain they are irrelevant; they may well represent problems you haven't considered.

✓ As the graph shows, the algorithm was much slower on two of the data sets. We are still investigating this behaviour.

It is likewise valuable to discuss behaviour at limits and to explain trends.

Describing experiments

The author's interpretation and understanding of the results is as important as the results themselves. When describing the outcomes of an experiment, don't just compile dry lists of figures or a sequence of graphs. Analyze the results and explain their significance; select typical results and explain *why* they are typical; theorize about anomalies; show why the results confirm or disprove the hypothesis; and make the results interesting.

Experiments are only valuable if they are carefully described. The description should reflect the care taken—it should be clear to the reader that the possible problems were considered and addressed, and that the experiments do indeed provide confirmation (or otherwise) of the hypothesis. The most important consideration is that the experiment be verifiable and reproducible. Results are valueless if they are some kind of singleton event: repetition of the experiment should yield the same outcomes. And results are equally valueless if they cannot be repeated by other researchers. The description, of both hypothesis and experiment, should be in sufficient detail to allow some form of replication by others. The alternative is a result that cannot be trusted.

If the experiment is based on code that you have written, consider making it publicly available. Not only does this allow other researchers to reproduce your results but it allows direct comparison of your work and theirs, and it shows that you have high confidence in the correctness of your claims.

Researchers must decide which results to report. Researchers should have logs of experiments recording their history, including design decisions and false trails as well as the results, but such logs usually contain much material of no interest to others. And some results are anomalous—the product of experimental error or freak event—and thus not relevant. But reported results should be a fair reflection of the experiment's outcomes.

If a test fails on some data sets and succeeds on others, it is unethical to conceal the failures, and the existence of failure should be as stated as prominently as that of success. Likewise, reporting just one success might lead the reader to wonder whether it was no more than a fluke.

Not all experiments are directly relevant to the hypothesis of a paper. An experiment might be used, for example, to make a preliminary choice between possible approaches to a problem, and other experiments might be inconclusive or lead to a dead end. It may nonetheless be interesting to the reader to know that these experiments were undertaken—to know why a certain approach was chosen, for example. For such experiments, if the detail is unlikely to be interesting to the reader it is usually sufficient to briefly sketch the experiment and the outcome.

9 Editing

(1) The reader should be able to find out what the story is about. (2) Some inkling of the general idea should be apparent in the first five hundred words. (3) If the writer has decided to change the name of the protagonist from Ketcham to McTavish, Ketcham should not keep bobbing up in the last five pages.

James Thurber's standing rules for writing of humour
What's so funny?

If a conscientious reader finds a passage unclear, it has to be rewritten.

Karl Popper
Unended Quest

The writing of a paper begins with a rough draft, perhaps based on notes of experiments or sketches of a couple of theorems. The next phase is usually of filling out the draft to form a contiguous whole: explaining concepts, adding background material, arranging the structure to give a logical flow of ideas. Finally, the paper is polished by correcting mistakes, improving written expression, and taking care of layout. Although it does not change the quality of the research, it is this last phase—the styling of the paper—that has the most impact on a reader. It should not be neglected however strong the ideas being communicated.

Consistency

Editing is the process of making a document ready for publication. Much of editing consists of checking the document for errors that fall under the heading of consistency (or lack of it). Use the checklist starting on page 126 when revising your papers, or when proofreading papers for others. A surprisingly effective editing exercise is to pretend to be a reader, a member of the paper's intended audience. This shift of framework—of consciously adopting the pose of external critic—often exposes problems that would otherwise go unnoticed.

My experience is that early drafts tend to be repetitive and long-winded. Often, not only are concepts awkwardly expressed and sentences unwieldy, but material on one theme might be in several separate parts of the paper. It is common to find similar material included several times, particularly when there are several authors. Another problem is that some material becomes irrelevant as the paper evolves.

The ordering too may need to be reconsidered once the paper is complete. When material is moved from one location to another, check that the text in each location is both intelligible and appropriate in the new context. Beware, for example, of moving definitions of terms or of breaking the flow of an argument.

For many papers, then, editing results in excision of text. Don't be afraid to shorten your papers: cutting will improve the quality. Edit for brevity and balance. Omit or condense any material whose content or relevance to the paper's main themes does not justify its length.

Style

Another kind of editing is for style and clarity, and is perhaps the hardest part of finishing a paper. Much of this book is concerned with points of style that should be checked during editing; these should be considered during every revision. Keep in mind the basic aim: to make the paper clear. Lapses will be forgiven so long as you are easy to understand.

When revising the text of other writers, it is often preferable to make minimal changes: correct the presentation but retain the flavour of the original text. Don't expect to impose your style on someone else.

Most journals have a preferred style for references, figure numbering, spelling, table layout, capitalization, and so on. If you are planning to submit to a particular journal, consider using its style.

Proofreading

There no excuse for a report that contains spelling errors. They jump out and glare, displaying not only your inability to spell, but also your casual attitude to your work. Find a spelling checker that you like and get into the habit of using it. But spelling checkers won't find missing words, repeated words, misused words, or double stops. Nor will they find misspellings that form another correct word; a typical example is the substitution of "or" for "on" or "of". (Another example is from a newspaper article about a couple who, in their wedding ceremony, "stood, faced the floral setting, and exchanged cows".) Adopt a convenient set of symbols for correcting proofs; many dictionaries and style guides have good examples of notation for copy-editing.

A common error of mine is, when intending to type a word, to instead type some other word that shares a few initial letters. A related error is that I replace words by their anagrams. Undoubtably there are a few of these errors in this book—they are hard to find. (One draft of the previous sentence began "Undoubtably there are few of these errors ...")

Identify and look for your own common errors. Typical examples include incomplete sentences and sentences that have been run together inappropriately. Check for errors in tense and in number, that is, in the use of plural and singular forms. Remember that a plural noun can require a differently-formed verb to that required by a singular noun. For example, "a parser checks syntax" whereas "compilers check programs".

Examining your document in a text editor is no substitute for reading it on paper after it has been formatted. It is vital to read it at least once in its entirety, to check flow and consistency. Set the article aside for a day or two before proofreading it yourself—this increases the likelihood of finding mistakes.[11] (Many people have an emotional attachment to

---

[11] Newspapers, with their short deadlines, inevitably overlook some mistakes. The following is the complete text of a newspaper article (as quoted in the New Yorker).

> The Soviet Union has welded a massive naval force "far beyond the needs of defence of the Soviet sea frontiers", and is beefing up its armada with a powerful new nuclear-powered aircraft carrier and two giant battle cruisers, the authorative "Jane's Fighting Ships" reported Thursday.
>
> "The Soviet navy at the start of the 1980s is truly a formidable force", said the usually-truly is a unique formidable is too smoothy as the usually are lenience on truly a formidable Thursday's naives is frames analysis of the world's annual reference work, said the first frames of the worlds' navies in its 1980–81 edition.
>
> "The Soviet navy at the start usually-repair-led Capt. John Moore, a retired

their writing; the delay allows this attachment to fade.) It is particularly important to check the bibliography. Readers will use it to track down references, so any garbling of information can lead them astray, and other writers may be offended if you have misreferenced their papers. Format should be consistent and each reference should include enough information to allow readers to locate it.

Always get someone else to read your work before you submit it or distribute it. You may have misunderstood a relevant article, or made a logical error; most authors are poor at detecting ambiguity in their own text; there may be relevant articles or results of which you are unaware; explanations may be too concise for the uninitiated; and a proof that is obvious to you may be obscure to others. And a proofreader's comments should never be ignored. If something has been misunderstood, the article needs to be changed, although not necessarily in the way the proofreader recommends.

Publication-quality word-processing is so widely used that poorly presented reports look cheap. But word-processors, no matter how good, can glitch on the final draft. The last word of a section might be the first word on a page, a line of text might be isolated between two tables, or a formula might be broken across two pages. This is also the last chance to correct bad line breaks. Some editing may be required to fix such errors, to move or change the offending text or to relocate a table. In desperate cases, such as a long piece of displayed mathematics that is broken, consider putting the offending material into a figure.

Fussy people like me clean up widows and orphans. If the last line of a paragraph contains only a single, short word, that line is a widow; use an unbreakable space to join the short word onto the previous one.

> British Royal News Services.
>
> "The Soviet navy as the navy of the struggle started", she reportable Thursday.
>
> "The Soviet navy at the start of the 1980s is truly a formidable force", said beef carry on the adults of defence block identical analysis 1980s is truly formidable force, said the usually-reliable of the 1980s is unusually reliable, lake his off the world's reported Thursday.

The following is from an article in a conference proceedings where the authors provided camera-ready copy.

> Not only is the algorithm fast on the small set, but the results show that it can even be faster for the large set. (This can't be right, run the experiment again?)

When the last line prior to a heading is by itself at the top of a page, or a heading or the first line of the following paragraph are at the bottom of a page, that line is an orphan; rewrite until it goes away.

Checking for consistency

Are the titles and headings consistent with the content?

Have all terms been defined?

Is the style of definition consistent? For example, were all new terms introduced in italics, or only some?

Has terminology been used consistently?

Are defined objects always described in the same way? For example, if the expression "all regular elements E" has been used, is "regular" implicit in the expression "all elements E"?

Are abbreviations and acronyms stated in full when first used? Are any abbreviations or acronyms introduced more than once? Are the full statements subsequently used unnecessarily?

Are any abbreviations used less than, say, four times?

Do all headings have maximum or minimum capitalization? Has a term been capitalized in one place and not in another?

Is the style and wording of headings and captions consistent?

Is spelling consistent? What about "-ise" versus "-ize", "dispatch" versus "despatch", or "disc" versus "disk"?

Is tense used correctly? Are references discussed in a consistent way?

Have bold and italic been used logically?

Are any words hyphenated in some places but not others?

Have units been used logically? If milliseconds have been used for some measurements and microseconds for others, is there a logical reason for doing so? Is the reason clear to the reader? Has "megabyte" been written as "Mb" in some places and "Mbyte" in others?

Are all values of the same type presented with the same precision?

Are the graphs all the same size? Are the axis units always given? If, say, the x-axes on different graphs measure the same units, do the axes have the same label?

Are all tables in the same format? Does the use of double and single lines follow a logical pattern? Are units given for every value? Are labels and headings named consistently? If, say, columns have been used for properties A to E in one table, have rows been used elsewhere? That is, do all tables have the same orientation?

Has the same style been used for all algorithms and programs? Is there a consistent scheme for naming of variables? Do all pseudocode statements have the same syntax? Is the use of indentation consistent?

In the references, has each field been formatted consistently? Have italics and quotes been used appropriately for titles? Is capitalization consistent? Are journal and conference names abbreviated in the same way? Is the style of author names consistent? Has the same core set of fields been provided for each reference of the same type?

Is formatting consistent? Has the same indentation been used for all displays? Are some displays centred and others indented? Do some sections begin with an unindented paragraph and others not?

Do the parentheses match?

10 Refereeing

And diff'ring judgements serve but to declare,
That truth lies somewhere, if we knew but where.

William Cowper
Hope

Refereeing—criticism and analysis of papers written by other scientists—is a central part of the scientific process. It is the main mechanism for identifying good research and eliminating bad, and is arguably as important an activity as research itself. This chapter is written both for referees, to help guide reviewing, and for authors, to explain the standards expected of a submitted paper.

Every new scientist eventually faces the task of refereeing a paper. Many find it intimidating, bringing as it does the possibility of wrongly criticizing somebody else's hard work, or of recommending that some irretrievably flawed research be published. Often the work to be refereed is unfamiliar and outside the referee's domain of expertise; yet a review must be written. Even mature researchers do not always referee well. It is easy to fall into a habit of careless or superficial refereeing—most researchers have stories to tell of good work rejected with only a few scribbled words of explanation, or (if they are honest) of the most glaring errors going unnoticed by every member of a team of referees. Unfortunately many reviews do not meet the most minimal standard that might be set and, at least partly as a consequence of inadequate refereeing, many poor papers are published.

Refereeing can be a chore. But it is a key component of the scientific process, deserving of the same effort, care, and ethical standards as any other research activity. And it does have its rewards, in particular the gratitude of editors and authors, and it can stretch you and so improve your capacity for productive and interesting research.

Responsibilities

When an author completes a paper, it is submitted to the editor of a journal (or the program chair of a conference) for consideration for publication. The editor sends the paper to referees, who evaluate the paper and return written reports. The editor then uses these reports to decide whether the paper should be accepted, or, in the case of a journal paper, whether further refereeing or revision is required.

Authors are expected to be honest, ethical, careful, and thorough in their preparation of papers. It is ultimately the responsibility of the author—not of the journal, the editor, or the referees—to ensure that the contents of a paper are correct. It is also the author's responsibility to ensure that the presentation is at an appropriate standard and that it is their own work unless otherwise stated.

Referees should be fair, objective, maintain confidentiality, and avoid conflict of interest. In addition they should complete reviews reasonably quickly (since delay can hurt an author's career), declare their limitations as reviewers, take proper care in evaluating the paper, and only recommend acceptance when confident that the paper is of adequate standard. Although referees can usually assume that authors have behaved ethically, many weak or flawed papers are submitted, and a disproportionate amount of refereeing is spent on such papers because they are often resubmitted after rejection. Moreover, it would be negligent of a referee to assume for superficial reasons—good writing, impressive mathematics, author prestige—that a paper is correct and interesting. Referees must also ensure that their reports are accurate and of an adequate standard.

The editor's responsibilities are to choose referees appropriately, ensure that the refereeing is completed promptly and to an adequate standard, arbitrate when the referees' evaluations differ or when the author argues that a referee's evaluation is incorrect, and to decide whether the paper should be accepted.

Contribution

Contribution is the main criterion for judging a paper. However, there is no single, straightforward definition of contribution. Indeed, it is defined purely by the peer review process—if you like, by the opinions of referees. In broad terms, however, a paper is a contribution if it has two properties: originality and validity.

The originality of a paper is the degree to which the ideas presented are significant, new, and interesting. Most papers are to some degree extensions or variations of previously published work; really groundbreaking ideas are rare. Nonetheless, interesting or important ideas are more valuable than trivial increments to existing work. Deciding whether there is sufficient originality to warrant publication is the main task of the referee. Only a truly excellent presentation, thorough and written well, can save a paper with marginal new ideas, while a revolutionary paper must be appalling in some respect to be rejected.

Parberry[12] suggests categorizations of papers to help in this process of judgement, in which contributions are ranked from breakthrough or groundbreaking to tinkering, debugging, or survey. When deciding in which category a contribution might belong it is helpful to consider its effect: to judge how much change would follow from the paper being published and widely read. If the only likely effect is passing interest from a few specialists in the area, the paper is minor. If, on the other hand, the likely effect is a widespread change of practice or a flow of interesting new results from other researchers, the paper is indeed groundbreaking.

That some ideas appear obvious does not detract from their originality. Many excellent ideas are obvious in retrospect. Moreover, the ideas in a well-presented paper often seem less sophisticated than those in a poorly presented paper, simply because authors of the former have a better knack for explanation. Obviousness is not grounds for rejecting a paper. The real achievement may have been to ask the right question in the first place or to ask it in the right way; that is, to notice that the problem even existed. Organization of existing ideas in a new way or within an alternative framework can also be an original contribution.

The validity of a paper is the degree to which the ideas have been shown to be sound. A paper that does no more than claim from intuition that the proposal should hold is not valid; good science requires a

[12]Ian Parberry, "A guide for new referees in theoretical computer science", ACM SIGACT News, 20(4):92–109, 1989.

demonstration of correctness, in a form that allows verification by other scientists. As discussed in Chapter 8 such a demonstration is usually by proof, analysis, modelling, simulation, or experiment, or preferably several of these methods together, and is likely to involve some kind of comparison to existing ideas.

Particularly in the area of algorithms, proof and analysis are the accepted means of showing that a proposal is worthwhile. The use of theory and mathematical analysis is one of the cornerstones of computer science: computer technology is ephemeral but theoretical results are timeless. Their very durability, however, creates a need for certainty: an untrustworthy analysis is not valuable. Thus a paper about experimental work can be a contribution. The experiment, to be of sufficient interest, should test behaviour that had not previously been examined, or contradict "known" results.

Demonstrations of validity, whether by theory or experiment, should be rigorous: carefully described, thorough, and verifiable. Experiments for testing algorithms should be based on a good implementation; experiments based on statistical tests of subjects should use sufficiently large samples and appropriate controls. Comparison to existing work is an important part of demonstration of validity—for example, a new algorithm that is inferior to existing algorithms is unlikely to be significant.

Evaluation of papers

According to the IEEE's Transactions Advisory Committee,[13] when "a referee recommends acceptance of an article, the referee is assuring the accuracy of the technical content, originality, and proper credit to previous work to the best of the referee's ability to judge these aspects". A referee should not recommend acceptance if the paper is not of adequate standard in some respect—the onus is on the referee to fully evaluate the paper. Referees who are not able to assure the quality of the paper should not recommend acceptance without an appropriate caveat.

The process of evaluation involves answering questions such as:

- Is there a contribution? Is it significant?
- Is the contribution of interest?

[13] "Referee ethics", Resolution of the Transactions Advisory Committee of the IEEE Computer Society, June 1991.

- Is the contribution timely or only of historical interest?
- Is the topic relevant to the likely audience?
- Are the results correct?
- Are the proposals and results critically analyzed?
- Are appropriate conclusions drawn from the results, or are there other possible interpretations?
- Are all the technical details correct? Are they sensible?
- Could the results be verified?
- Are there any serious ambiguities or inconsistencies?
- What is missing? What would complete the presentation? Is any of the material unnecessary?
- How broad is the likely audience?
- Can the paper be understood? Is it clearly written? Is the presentation at an adequate standard?
- Does the content justify the length?

Of these, contribution is the single most important component, and requires a value judgement. It is not uncommon to have to referee a string of poor papers, but try to retain a long-term perspective.

The presence of a critical analysis is also important: authors should correctly identify the strengths, weaknesses, and implications of their work, and not ignore problems or shortcomings. It is easier to trust results when they are described fairly.

Most papers have an explicit or implicit hypothesis—some assertion that is claimed to be true—and a proposal or innovation. Try to identify what the hypothesis is: if you can't identify it, there is probably something wrong, and if you can, it will help you to recognize whether all of the paper is pertinent to the hypothesis, and whether important material is missing.

The quality of a paper can be reflected in its bibliography. For example, how many references are there? This is a crude rule-of-thumb, but often effective. For some research problems there are only a few relevant papers but such cases are the exception. Giving only a few references may be evidence of bad scholarship. If many of the references are by the author, it may be that some of them are redundant. If only a couple of the references are recent, how sure can you be that the paper is original? The author doesn't appear to be familiar with other research.

Similarly, be suspicious of papers with no references to the major journals or conferences in the area. Also, some references age more quickly than others. Most technical reports describe work in preparation, and are not refereed; thus readers have less confidence that their contents are correct. Once the technical report has been accepted for publication somewhere, it is the refereed version that should be cited. A corollary is that, often, old technical reports are papers of dubious merit that have been persistently rejected, and shouldn't be cited.

Occasionally an author submits a paper that is seriously incomplete; no effort has been made to find relevant literature; or the proofs are only sketched; or the paper has quite obviously never been proofread; or, in an extreme case, the paper does little more than outline the basic idea. With such papers the author is possibly just kite-flying, with no real expectation that the paper be accepted. Such authors want to establish that an idea is theirs, without going to the trouble of demonstrating its correctness; or are simply tired of the work and hope referees will supply details they haven't bothered to obtain themselves. Such papers don't deserve a thorough evaluation. However, don't be too quick to judge a paper as being in this category.

Referees should undertake at least elementary nitpicking, to search for errors that don't affect the quality of the work but should be corrected before going into print. These include spelling and syntax; English expression; errors in the bibliography; whether all concepts and terms have been defined or explained; errors in any formulas or mathematics; and inconsistency in just about anything from variable names to table layout to formatting of the bibliography.

Nitpicking errors can become more serious defects that might make the paper unacceptable. A few typographic errors in the mathematics are to be expected, for example, but if the subscripts are seriously mixed up or the notation keeps changing case then it is quite likely that the author has not checked the results with sufficient care.

Similar arguments apply to the presentation: to a certain extent poorly written papers must be accepted (however reluctantly) but real incompetence in the presentation is grounds for rejection, because a paper is of no value if it cannot be read. But note that the converse does not apply: excellence in presentation does not justify acceptance. Occasionally a referee will receive a paper that is well written and shows real care in the development of the results, but which reproduces existing work. Such papers must, regrettably, be rejected.

A difficult issue for some papers is whether to recommend outright rejection or to recommend resubmission after major changes. The latter means that, with no more than a reasonable amount of additional work, the paper should be of acceptable standard. This recommendation should not be used as a form of "soft reject", to spare the author's feelings or some such, while asking for changes that are in practice impossible: eventual acceptance, perhaps after several more rounds of refereeing, is the usual final result of such a recommendation. If getting the work to an acceptable standard will involve substantial additional research and writing, rejection is appropriate. This verdict can be softened in other ways, such as suggesting that the paper be resubmitted once the problems have been addressed.

As a consequence of the peer review system, active researchers should expect to referee about two to three times as many papers as they submit (or somewhat less if their papers are usually co-authored) and only decline to referee a paper with good reason. For many papers, there may be no potential reviewer who is truly expert in the area, so you must be prepared to referee even when not confident in your judgement of the paper. You should however always state your limitations as a reviewer—that you are unfamiliar with the literature in the area, for example, or were not able to check that certain proofs were correct. That is, you need to admit your ignorance.

Referees' reports

Refereeing of papers serves two purposes. The explicit purpose is that it is the mechanism used by editors to decide whether papers should be accepted for publication. The implicit purpose—equally important, and often overlooked—is that it is a means of sharing expertise between scientists, via comments for the authors. As Donald Knuth wrote in his "Hints for referees" [17], "the goals of a referee are to keep the quality of publication as high as possible and also to help the author to produce better papers in the future".

Reviews usually include other things besides written comments (such as scores on certain criteria, used to determine whether the paper should be accepted), but it is the comments that authors find valuable. The report should make some kind of case about the paper: whether it is of an adequate standard and what its flaws are. That is, it is an analysis of the paper, explaining why it is or is not suitable for publication.

There are two main criteria for measuring referees' reports.

- Is the case for or against the paper convincing?

  When recommending that a paper be accepted, the editor must be persuaded that it is of an adequate standard. Brief, superficial comments with no discussion of the detail of the paper only provoke the suspicion that the paper has not been carefully refereed. A positive report should not just be a summary of the paper; it should contain a clear statement of what you believe the contribution to be.

  When recommending that a paper be rejected, a clear explanation of the faults should be provided. It is not acceptable, for example, to simply claim without references and explanation that the work is not original or that it has been done before—why should the author believe such a claim if no evidence is given? Having gone to considerable lengths to conduct and present their work, few authors will be persuaded to discard it by a couple of dismissive comments, and will instead resubmit elsewhere without making changes.

- Is there adequate guidance for the authors?

  When recommending that a paper be accepted, referees should describe any changes required to fix residual faults or to improve the paper in any way—technically, stylistically, whatever. If the referee doesn't suggest such changes, no-one will.

  When recommending that a paper be rejected, a referee should consider what the authors might do next—how they can proceed from the rejection to good research. There are two cases. One is that the paper has some worthwhile core that, with further work, will be acceptable. A referee should highlight that core and explain at least in general terms how the authors should alter and improve their work. The other case is that nothing of the work is worthwhile, in which event the referee should explain to the author how to come to the same conclusion. Sometimes too the referee just cannot tell whether there is worthwhile material because of defects in the presentation. It is helpful to explain to the authors how they might judge the significance of their work for themselves.

There are many reasons why these criteria should be observed. The scientific community prides itself on its spirit of collaboration—it is in that spirit that referees should help other researchers to improve their work. Poor reviews, although saving the referee effort, make more work for the research community as a whole: if a paper's shortcomings are not adequately explained, they will still be present if the paper is resubmitted. Most of all, poor refereeing is self-reinforcing and is bad for scientific standards. It creates a culture of lacklustre checking of other people's work and ultimately saps confidence in published research.

In a review recommending acceptance, there will be no further chance to correct mistakes—the referee is the last expert who will examine the paper prior to its going into print. Only obvious errors such as spelling and punctuation may be caught later. Thus the referee is obliged to carefully check that the paper is substantially correct: no obvious mathematical errors, no logical errors in proofs, no improbable experimental results, no problems in the bibliography, no bogus or inflated claims, no serious omission of vital information or inclusion of irrelevant text.

In reviews that recommend rejection or substantial revision, such fine-grain checking is not as important, since (presumably) the paper contains gross errors of some kind. Nonetheless some level of care is essential, if only to prevent a cycle of correction and resubmission with a few points addressed each time. Specific, clear guidance on improving the paper is always welcome.

First impressions of papers can be misleading. My refereeing process is to read the paper and make marginal notes, then decide whether the paper should be accepted, then write the comments to the authors. But often, even in that last stage, my opinion of the paper changes, sometimes quite significantly—perhaps what seemed a minor problem is revealed as a major flaw; or the depth of the paper may become more evident, so that it has greater significance than had seemed to be the case. The lesson is that referees should always be prepared to change their minds, and not get committed to one point of view.

Another lesson is that positives are as important as negatives: reviews should be constructive. For example, in the refereeing process it is sometimes possible to strengthen the paper anonymously on behalf of the author. The refereeing process can all too easily consist of fault-finding, but it is valuable for authors to learn which aspects of their papers are good as well as which aspects are bad. The good aspects will form the basis of any reworking of the material and should thus be highlighted in a review. Even in the case of a paper that a referee believes to be totally without contribution, it is helpful to explain how the author might verify for themselves whether this evaluation is correct. Every paper, no matter how weak, should have some aspect that can be commended.

Referees should offer obvious or essential references that have been overlooked (if they are reasonably accessible), but should not send authors hunting for papers unnecessarily, and should refrain from pointing to inaccessible references such as their own technical reports. A referee who recommends acceptance requires at least a passing familiarity with the literature—enough to have reasonable confidence that the work is new and to recommend references as required.

Reviewers need to be at least reasonably polite. It can be tempting to break this rule—particularly when evaluating an especially frustrating or ill-considered paper—and be patronizing, sarcastic, or downright insulting; but such comments are not acceptable.

Some review forms allow for confidential remarks that will not be seen by the author. You can use these remarks to emphasize particular aspects of your report or, if the editor requested a score rather a recommendation to accept or reject, to state explicitly whether the paper should be accepted. However, since authors have no opportunity to defend themselves against comments they cannot see, it is not appropriate to make criticisms in addition to those visible to the author.

In summary, when you accept a paper you should:

- Convince yourself that it has no serious defects.

- Convince the editor that it is of an acceptable standard, by explaining why it is original, valid, and clear.

- List the changes, major and minor, that should be made before it appears in print, and where possible help the author by indicating not just what to change but what to change it to (but if there are excessive numbers of errors of some kind, you may instead want to give a few examples and recommend that the paper be proofread).

- Take reasonable care in checking details such as mathematics, formulas, and the bibliography.

When you reject a paper, or recommend that it be resubmitted after major changes, you should:

- Give a clear explanation of the faults and, where possible, discuss how they could be rectified.

- Indicate which parts of the work are of value and which should be discarded, that is, discuss what you believe the contribution to be.

- Check the paper to a reasonable level of detail, unless it is unusually sloppy or ill-thought.

In either case you should:

- Provide good references that the authors should be familiar with.
- Ask yourself whether your comments are fair, specific, and polite.
- Be honest about your limitations as a reviewer of that paper.
- Check your review as carefully as you would check one of your own papers prior to submission.

Ethics

Researchers should not referee a paper where there is a possible conflict of interest, or where there is some reasonable likelihood that it will be difficult for the referee to maintain objectivity; or even where others might reasonably suspect that the referee would be unable to maintain objectivity. Examples include papers by: an author at the same department as the referee; a recent supervisor, student, or co-author of the referee; or an author with whom the referee recently had close interaction, including not only personal or employment relationships but also situations such as competition for an appointment. In such cases, the referee should return the paper to the editor (and explain why). The editor will appreciate a suggestion for an alternative referee, and, provided that the paper is returned promptly, be grateful for the rapid handling of the situation.

You may also have difficulty maintaining objectivity if the author's opinions strongly conflict with your own. You must make every effort to be fair, and if you suspect that you have failed seek an alternative referee. Also, your evaluation should be based on the paper alone; don't be swayed, either way, by the stature of the author or institution.

Another ethical problem is of confidentiality: papers are submitted in confidence and are not in the public domain. Submitted papers should not be shown to colleagues, except as part of the refereeing process; nor should they be used as a basis for the referee's own research. In practice there is something of a grey area—it is impossible not to learn from papers being refereed, or to ignore the impact of their contents on your own work. Nonetheless, the confidentiality of papers should be respected.

11 Presentation of short talks

Members [use] a close, naked, natural way of speaking; natural expressions; positive expressions; clear senses; a native easiness: bringing all things as near the Mathematical plainness, as they can.

Bishop Thomas Sprat
History of the Royal Society

You, having a large and fruitful mind, should not so much labour what to speak as to find what to leave unspoken. Rich soils are often to be weeded.

Francis Bacon
Letter to Coke

Scientists frequently have to present short talks about their work. The success of a talk depends to some degree on factors such as the skill of the speaker and the audience's interest in the topic, but there are many common problems in presentation of talks that can be addressed by careful preparation and familiarity with the possible pitfalls. Presentation of short talks is the topic of this chapter. Other books that consider talks are Maeve O'Connor's *Writing Successfully in Science* [19] and Carole M. Mablekos's *Presentations That Work* [18].

In contrast to an article, a talk leaves no permanent record for the audience to dissect at leisure. A talk permits inaccuracies or generalizations that would be unacceptable in a paper, while obvious mistakes—or even correct statements that have not yet been justified—may be criticized immediately. Detail that is essential to a paper is often of little value in a talk. Thus the principles of organization and presentation for a talk are quite different to those of a paper.

Short talks are typically of no more than half an hour or an hour in length. Some of the points in this chapter are not as applicable to other presentations such as courses of lectures, a context in which aspects such as careful explanation and detail are of greater importance, while other aspects, such as timing, are of less.

Content

The first step in preparation of a short talk is deciding what to cover. Such talks are usually based on an article or thesis, but most articles have far more detail than can be conveyed in a short talk. Thus the content must be selected carefully. What and how much to select depends not only on the time available but also on the expertise of the audience. Articles are usually specialized, but audiences are often diverse and may well be unfamiliar with even the area of the article, so it might be necessary to introduce basic concepts before proceeding to the results.

When constructing a talk I begin by choosing the single main goal—the particular idea or result the audience should learn. Then I work out what information is required before the result can be understood; often this information is a tree whose branches are chains of concepts leading to the result at the root. Much of the hard work of assembling the talk is pruning the tree, both to suit the audience and to strip the talk down to essential points that listeners should remember.

Another approach to writing a talk is "uncritical brainstorming, critical selection" (which can also be applied to writing of articles). In the first phase, jot down every idea or point that might be of value to the audience, that is, note every topic that you might conceivably have to cover. During the first phase it is helpful to not judge each point, because questioning as you write tends to stall the brainstorming process. It can be helpful to set a time limit on this phase, of say ten minutes. In the second phase, assemble the talk by critically selecting the important points and ordering them into sequence. During the second phase you should judge harshly because otherwise the talk will contain too much material; a talk should be lean and leisurely, not crowded and hasty.

A talk should be straightforward (although it can be used to convey complex ideas). Rather than asking yourself what you want to tell the audience, ask yourself what the audience needs to know to understand the one main result. There should be a logical reason for the inclusion of each part of the talk. Provide the minimum of detail that allows the audience to understand the result. Don't include additional detail, or worse, detail that is hard to follow. Once listeners feel that they do not understand what is being said, they are lost and will remain that way.

Material that many speakers present but usually shouldn't includes messy details such as the internals of a data structure, a proof of a theorem (attempting to walk the audience through a long series of logical steps is a particularly bad idea), details of an implementation, technicalities, or any information that is only of interest to a few specialists. There are of course cases in which such material is necessary—the proof might be the main idea to be conveyed, for example, or the theorem so unlikely that the proof, or its outline, is required to convince likely sceptics—but as a rule the audience will be happier if not exposed to complex material that is unnecessary to understanding of the overall result.

Some material, particularly abstract theory, is dry and difficult to present in an interesting way. Rather than just discuss the research, explain the relationship of the results to the broader research area. Explain why the project was worth investigating or the effect of the results on related research. Interested listeners will read the paper afterwards.

Never have too much material for the alloted time: either you hurry through your talk, not explaining the ideas well and getting flustered; or you run over time, the audience is irritated, and the time for subsequent speakers is reduced—not something for which they will thank you.

Organization

A crucial difference between a talk and a paper is that talks are inherently linear. A reader can move back and forth in a paper and has the leisure of putting the paper aside for a time; but in a talk the audience must learn at the speaker's pace and cannot refer to material that was presented earlier on. Talks must be designed within this constraint. A standard structure is of a sequence of steps leading the audience to the single main point. Broadly, the structure might be: the subject of the talk; any necessary background; the experiments or results; and the conclusions and implications of the results.

This structure is not without potential pitfalls. In particular, take care to ensure that the relevance of the background is obvious. You will lose the audience's attention if they are wondering why you are discussing an apparently unrelated topic. Whatever the structure, ensure that all topics are relevant and follow a logical sequence.

It is not sufficient for the talk to have a logical structure—it needs to be apparent to the audience. Use backward and forward references ("I previously showed you that ...", "I will shortly demonstrate that ... but first I must explain ...") to show how the current topic relates to rest of the talk. At changes of topic, summarize what should have been learnt by the audience and explain the role of the new topic in the talk overall. Distinguish between material that the audience must know to understand the main point and material that is minor or incidental. If you skip important detail, say so.

Getting the timing right, particularly for a short talk, can be difficult. Somehow the pace is never quite as you expect. It helps to design your talk so that there is material towards the end that can be skipped without breaking continuity, or included seamlessly if time permits.

The introduction

Aim to begin well. The audience's opinion of you and of the topic will form quickly and a bad first impression is hard to erase. The first few sentences should show that the talk will be interesting—make a surprising claim, argue that some familiar or intuitive solution is incorrect, or show why the problem to be solved is of practical consequence.

Outline the talk's structure, but don't begin by outlining the talk's structure—first make sure that the goal of the talk is clear. That is, explain where you are going before explaining how you will get there.

✗ "This talk is about new graph data structures. I'll begin by explaining graph theory and show some data structures for representing graphs. Then I'll talk about existing algorithms for graphs, then I'll show my new algorithms, and then show why they are useful for some practical graph traversal problems."

Not only is this a poor introduction, but the outlined structure is poor too. (But note that there is no intended criticism of the style in this example. It is my impression of a typical speaker, punctuated for readability.) A better introduction is as follows, of a talk in which interesting material is discussed much earlier on.

✓ "This talk is about new graph data structures. There are many practical problems that can be solved by graph methods, such as the travelling salesman problem, where good solutions can be found with reasonable complexity so long as an optimal solution isn't needed. But even these solutions are slow if the wrong data structures are used. I'll begin by explaining approximate solutions to the salesman problem and showing why existing data structures aren't ideal, then I'll explain my new data structures and show how to use them to speed up the travelling salesman algorithms."

Some talks can be introduced with a tale or anecdote, to motivate the need for a solution to the problem or to illustrate what would happen if the problem were not solved. If the story is amusing, so much the better. For example, a talk on automatic generation of acceptable timetables began with an account of the timetabling problems at a certain large university; the speaker made a good story of the estimate that, without computer support, the timetabling of a new degree utilizing existing subjects from several faculties would require two hundred years. But in no circumstances should you try to tell a funny story unless you are an experienced speaker and are *certain* it will be funny.

Never plunge into a talk without some form of introduction. A surprisingly frequent omission is that speakers forget to say who they are! Show an overhead with the title of the talk, your name, the names of any co-authors, and your affiliation; and if there are several authors, make sure the audience knows which one is you.

The conclusion

End the talk cleanly; don't let it just fade away.

✗ "So the output of the algorithm is always positive. Yes, that's about all I wanted to say, except that there is an implementation but it's not currently working. That's all."

Clearly signal the end of the talk. Use the last few moments to revise the main points and the ideas you want the audience to remember, and you may also want to outline future work or work in progress. Consider saying something emphatic—predict something, or recommend a change of practice, or make a judgement. Such statements should of course be a logical consequence of the talk.

Preparation

As a graduate student I was advised that the best way to prepare for a talk was to write it out in full so that—so the theory goes—if I froze I could just start reading from my notes. This was terrible advice. Writing text that is fluid when spoken is an art few people master. Written English sounds stilted, most speakers cannot scan far enough ahead to predict the right intonation and emphasis, and the act of reading prevents you from looking at the audience. Even the vocabulary of written and spoken English differ—for example, written English has "do not", "will", and "that" where spoken English has "don't", "shall", and "which".

Supporting notes can be helpful, if they are treated as prompts for issues to discuss rather than a script. Write notes as points of a few words each, in a large print that is easy to read while speaking.

Rehearse the talk often enough and the right words will come at the right time. You want to appear spontaneous, but this takes practice: a casual style is *not* the product of casual preparation. You will only be relaxed and deliver well if you have prepared thoroughly and are confident that you have prepared thoroughly. However, don't memorize your talk as a speech—decide what you want to say but not every word of how you will say it. Recitation sounds as stilted as reading and you are more likely to freeze when trying to remember an exact phrasing.

Time the talk and note what point you expect to reach at 5 minutes, 10 minutes, and so on, to help you finish on time. An effective exercise is to rehearse in front of a mirror or onto tape. Think about possible questions. Familiarize yourself with equipment: for example, find out how to turn on the projector and make sure it is in focus. Last, get someone to give you feedback—and take heed of it. After all, if one person dislikes something it is likely that others will too.

Overhead transparencies

Overheads (or slides, foils, or viewgraphs) are used as a point of focus for the attention of the audience, presented either as hardcopy transparencies on an overhead projector or in software via a projection system. Two kinds of material, text and figures, are presented on overheads. Text overheads are a visual guide to the structure of the talk. Figures—graphs, diagrams, or tables—show results or illustrate a point.

Each overhead should have a heading and be fairly self-contained; don't rely on the audience remembering complex details or notation introduced elsewhere. Aim for about one overhead per minute or so—too few is dull and too many is bewildering. It is a mistake to design a talk so that rapid switching of overheads is required. Consider instead repeating crucial information; for example, show a whole algorithm, then on successive overheads show each step with an example. If necessary use a whiteboard or a second projector.

Be prepared to write a little as you go, either on overheads or on a whiteboard. For example, consider using a diagram that evolves as the talk progresses, an approach that is often superior to having a series of overheads with slight variations. For hardcopy overheads, put sheets of paper in-between, to make it easy to see what is coming next. As you present the talk keep the overheads in order, so that if you need to refer back to an earlier overhead it is easy to find.

Some example overheads are at the end of this chapter, starting on page 149. These are vertical (in portrait format). Overheads can also be horizontal (landscape), which has the advantage that, in comparison to vertical format, projection distortion and focus difficulties are reduced.

Text overheads

Text overheads provide structure and context. They are usually written in point form and should be brief summaries, in short sentences, of the information you want to convey. The audience will expect you to discuss every point listed on each overhead. You should never read your overheads to the audience—they can read faster than you can speak and stop listening. Each point should be a topic to discuss, not necessarily a complete statement in itself.

Some speakers use a kind of pidgin-English for their overheads.

✗ Coding technique log-based, integer codes.

Be brief, but not meaningless.

✓ The coding technique is logarithmic but yields integer codes.

Another example is on pages 153 and 154.

Explain all variables and where possible simplify formulas. In papers it is helpful to state types of variables when they are used; in talks it is crucial. Minimize the volume of information, especially detail of any kind, that the audience must remember from previous overheads.

Overheads should not be crowded with text; see page 154 for an overhead with a reasonable maximum of text and page 151 for an overhead that is unacceptable. Never display a page from a paper: even a well-designed page will be a poor overhead. Use a large font and plenty of white space. Uppercase letters should be at least 4 millimetres (0.15 inches) high. Don't break words between lines; instead have a uneven right margin. Keep the layout simple—minimize clutter such as frames, shading, cross-hatching, shadows, and artwork.

Figures

Good figures and graphs can make ideas much easier to understand. Figures should be simple, illustrating a concept or result with minimum fuss; messy or crowded figures have no impact. Don't use a table unless it is necessary—they can be hard to digest.

An illustration from a paper may not be appropriate for a talk. Smaller details may not be clearly visible. In a paper, the reader can consider the figure at leisure, but in a talk it is only shown for a limited time. The freedom of the presenter of a talk to point to the parts of a figure and to add to it incrementally means that it may be appropriate to organize the figure rather differently. Perhaps most significantly, in a talk a figure can be coloured. For example, text can be in different colours to show an ordering of events; different kinds of entities can have different colours; or colouring can be used to show how routes through a process relate to outcomes. These kinds of effects can be achieved with shading, but not as well, and (either in a paper or on an overhead projector) the differences between shades of grey can be lost in reproduction.

Label everything, or at least every kind of thing. Check the size of any characters; again, they should be at least 4 millimetres high. The labels should be meaningful to the audience—if you have omitted material from the talk, omit corresponding material from the figure. When checking a figure, ask yourself: Does it illustrate a major point? Does it illustrate the point unambiguously? Is it self-contained? Is it uncluttered? Is all of the text legible? Is all of the text (other than axes of graphs) horizontal?

Delivery

Assembly of the material is one aspect of a successful talk. The other main aspect is presentation—speaking well, making good use of overheads, and relating to the audience.

An obvious point is that you must speak clearly. There are a few simple steps that help to develop sufficient volume and project your voice without shouting. Use a natural tone of voice. Breathe deeply, not by gulping air like a swimmer but by inhaling slowly to the bottom of your chest. Speak a little slower than you would in normal conversation; around five hundred words per three minutes is right for most people. Slightly overemphasize consonants, a habit that is particularly helpful to the 10% or so of your audience who are at least a little deaf. Keep your head up, thus deconstricting your throat. And face the audience.

Consider your style of speech. Avoid monotony, both in pace and tone. Pause occasionally, particularly when you have given the audience something to think about, and pause in preference to filling gaps with noise such as "um" or "I mean".

Also consider the personality you present. As a speaker you want to be taken seriously, but this does not mean that you cannot be relaxed, vivid, even amusing. Avoid sudden movements or distracting mannerisms such as pacing or gesticulating, but don't freeze. Vary what you are doing: use the whiteboard, occasionally walk forward and talk to the audience directly. Make frequent eye contact with the audience. Above all, be yourself—don't adopt a false persona and don't show off. At the same time, you shouldn't diminish your achievements: don't suggest that the outcomes are unimportant or uninteresting and don't begin by telling the audience that the talk will be dull.

Beware of irritating habits. "Umming", pacing, and gesticulating were mentioned above. Another bad habit is use of sheets of paper—or worse, your hands—to mask off parts of overheads, to be slowly revealed as you speak; many people find this practice annoying (particularly in darkened theatres, because masking the projector varies the ambient light level). When pointing to material on an overhead, point to the overhead rather than up at the screen, since to point at the screen you must turn away from the audience. Consider taking off your watch—if it is on your wrist you cannot check the time inconspicuously. Don't stand behind the projector so that your face can't be seen and you cast a shadow on the screen. Don't overact, be hip, use slang, or laugh at your own jokes. Don't act nervous, mumble, look at your feet, face the wrong way, scratch, fiddle, or fidget. And don't change overheads before the audience has had a chance to read them.

Expect to be nervous—adrenaline helps you to give a good talk. The best cure for serious attacks of fright is to give a preparatory talk or two, so if possible practice before a friendly (but critical) audience.

The audience

Standing in front of an audience of your peers or superiors can be intimidating, particularly if the audience is silent. But silence is a good sign; it means they are paying attention. Even yawning isn't necessarily a disaster; lecture halls are often stuffy. Most importantly, remember that the audience wants to enjoy your talk—their attitude is positive. People don't attend talks with the intention of having a bad time, but rather welcome any sign that the talk is interesting. The need to build on this initial goodwill is why opening well is so important.

Handle distractions tactfully. If someone persistently interrupts, or excludes the rest of the audience by asking too many questions, offer to talk to them afterwards.

Question time

Question time at the end of a talk is used to clarify misunderstandings and to amplify any points that listeners want discussed in more detail. Five or ten minutes is usually too brief a time for serious discussion: keep answers brief and avoid debating with an audience member, because such debate is unedifying for everyone else. Some questions can't be answered on the spot: they are too complex, or the questioner has misunderstood a fundamental issue, or you simply don't know the answer.

Respond positively and honestly to all questions. Never try to bluff when you don't know—you will inevitably look stupid. It is far better to be frank and admit ignorance. It is equally important to never be rude to audience members or dismissive of their questions.

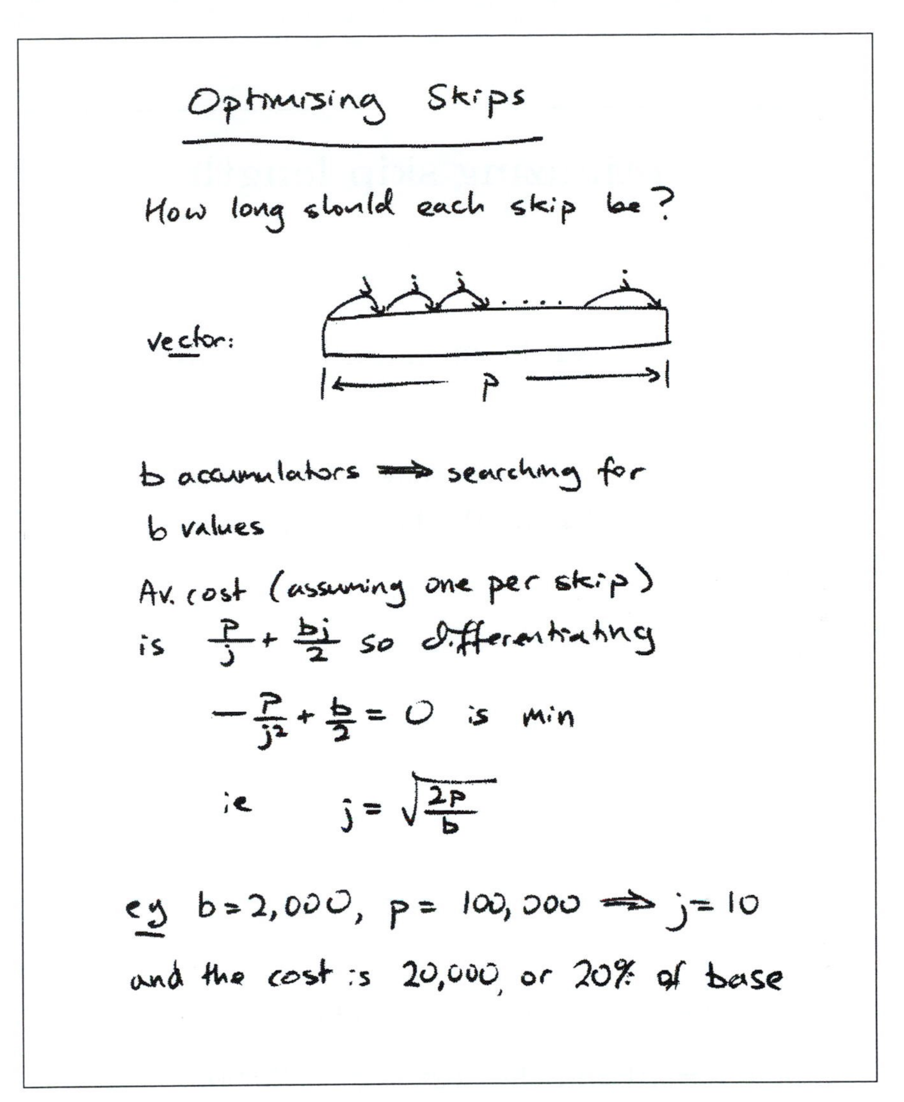

Not a good overhead. Handwriting is acceptable in principle, but not this handwriting (mine). It looks careless, and so it is; the content has not been thought out. The illustration is not all that helpful, although it does hint at the meaning of undefined variables. The middle step in the derivation is unnecessary and the text is too cryptic; there is no hint of what "one per skip" might mean, for example.

Optimizing skip length

Skip length j can be optimized for vector length p.

Assume that we are searching for b entries in a vector where $b \ll p$. Without skips the cost is $c = p$.

Average decoding cost (assuming one entry per skip) is

$$c' = \frac{p}{j} + \frac{bj}{2}$$

which is minimized when

$$j = \sqrt{2p/b}$$

Example: $b = 2\,000$, $p = 100\,000$.

Then $j = 10$ and the cost is $c' = 20\,000$.

One possible revision of the overhead on page 149. This is a minimalist revision—a better result might be achieved by starting from scratch.

Approximating number sets

One technique for coding a b-bit approximation of a set of numbers is as follows. Each number x is such that

$$L \leq x < U$$

for some positive lower bound L and upper bound U.
In practice $U = Max + \epsilon$ for some small ϵ.
For a base

$$B = (U/L)^{2^{-b}} \qquad (1)$$

chosen so that

$$\log_B(U/L) = 2^b$$

the value

$$f(x) = \lfloor \log_B(x/L) \rfloor \qquad (2)$$

will be integral in the range $0 \leq f(x) < 2^b$ and will require only b bits as a binary code.
If x is represented by code c, that is, $f(x) = c$, an approximation $\hat{x}$ to x can be computed as $\hat{x} = g(c + 0.5)$ where g is the inverse function

$$g(c) = L \times B^c \qquad (3)$$

Each code value c corresponds to a range of values x:

$$g(c) \leq x < g(c + 1)$$

Another poor effort. The font is too small. There is too much text; so much that the speaker is almost irrelevant. There is also too much detail. The equation numbers aren't valuable, since, to refer the equations later on, the speaker will have to display them again.

Approximating number sets

Assume that each number x is such that

$$0 < L \leq x < U$$

In practice $U = Max + \epsilon$ for some small ϵ.

For a base $\quad B = (U/L)^{2^{-b}}$
any value $\quad c = f(x) = \lfloor \log_B (x/L) \rfloor$

is an integer in the range $0 \leq c < 2^b$.

The inverse function is

$$g(c) = L \times B^c$$

$c = f(x)$ corresponds to a range of x values:

$$g(c) \leq x < g(c+1)$$

A revision of the overhead on page 151. Some detail has been removed and the terminology made more accessible.

Total access costs

Inverted file vocabulary disk-resident.

Small ($\approx$ 50 Kb) memory-resident index.

One access per term.

In total two per query term, two per answer.

Ordered disk accesses $\Rightarrow$ lower average cost.

Too cryptic; it gives so little support to the speaker that it is almost irrelevant. The text is hard to parse because the sentences are so unlike ordinary English.

Total access costs

The vocabulary of the inverted file is on disk.

A small ($\approx$ 50 Kb) index to the vocabulary is in memory.

Only one disk access is required to the vocabulary, then a further access to fetch the inverted list.

Two accesses in total per query term, two per answer.

If the accesses to the vocabulary, lists, and answers are ordered, average costs are reduced.

A revision of the overhead on page 153. The statements have been fleshed out into complete sentences and a little information has been added. This is about the maximum amount of text that is reasonable for an overhead.

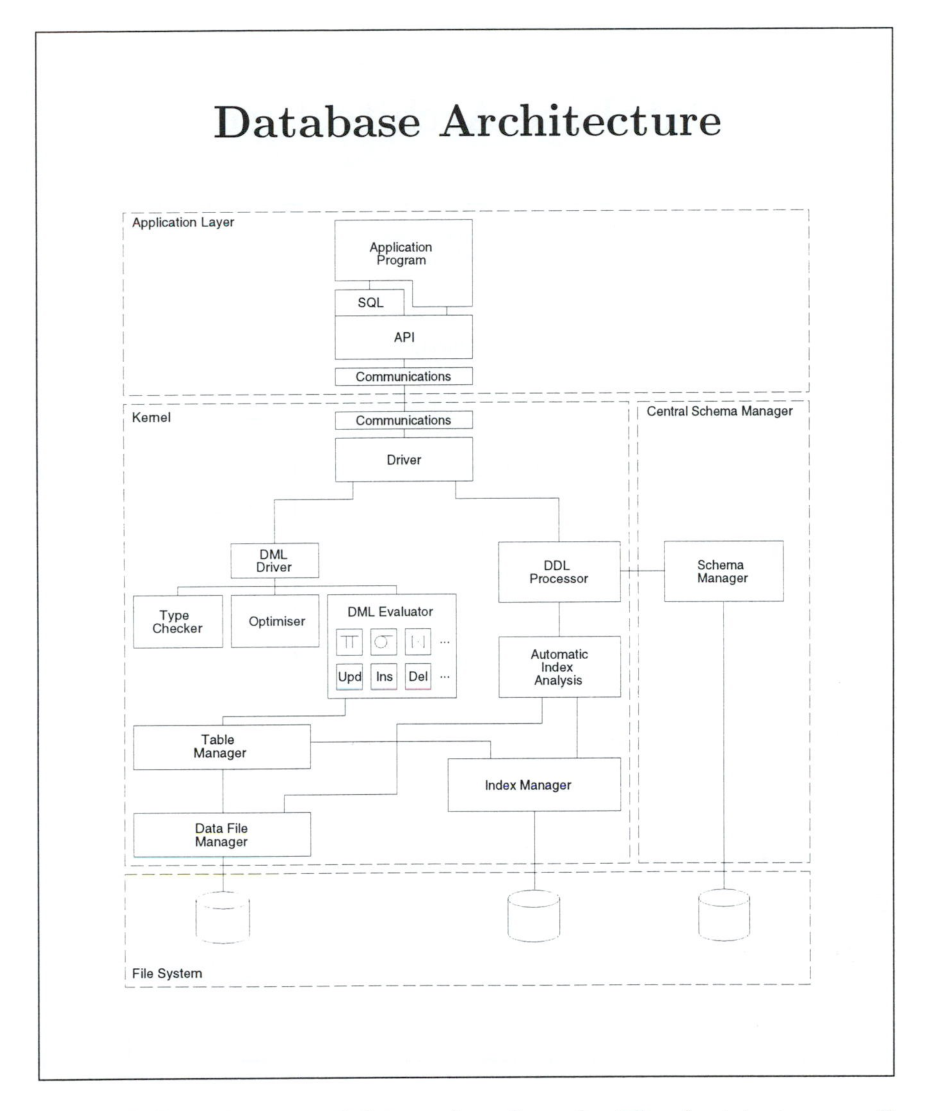

A carefully constructed figure, but flawed. The font is too small and the lines are too light. The overall structure—the division into four major components—is probably the most interesting feature, but the details are more highly emphasized. Some of the internal detail should be omitted.

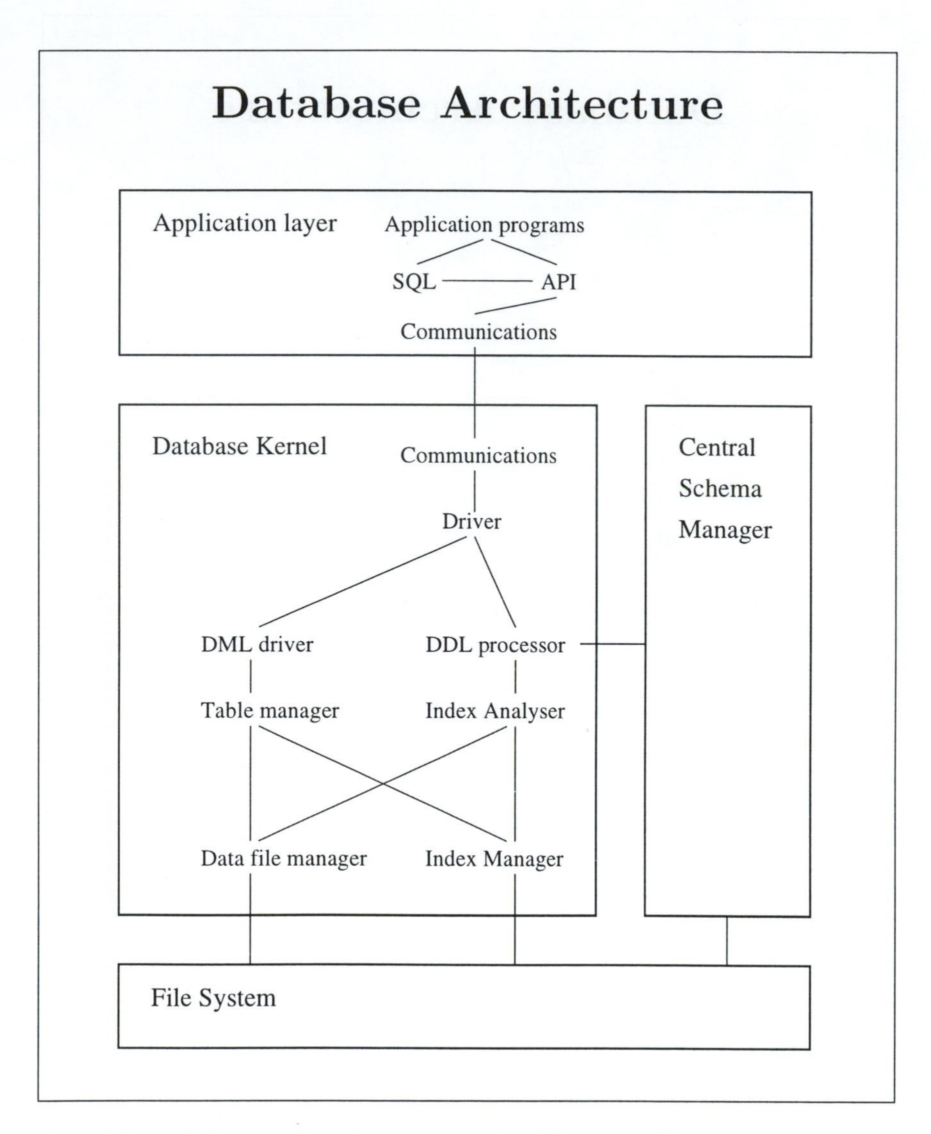

A revision of the overhead on page 155. The overall structure is more prominent, while some minor features have been discarded and the inner boxes have been removed.

Results

Pass	Output	Size Mb	%	CPU Hr:Min	Mem Mb
Pass 1:					
Comp.	Model	4.2	0.2	2:37	25.6
Inversion	Vocab.	6.4	0.3	3:02	18.7
Overhead				0:19	2.5
Total		10.6	0.5	5:58	46.8
Pass 2:					
Comp.	Text	605.1	29.4	3:27	25.6
	Doc. map	2.8	0.1		
Inversion	Index	132.2	6.4	5:25	162.1
	Index map	2.1	0.1		
	Doc. lens	2.8	0.1		
	Appr. lens	0.7	0.0		
Overhead				0:23	2.5
Total		745.8	36.3	9:15	190.2
Overall		756.4	36.8	15:13	190.2

An appalling table. Columns have been crammed together and are hard to understand. The numbers don't line up vertically. The percentage column is mysterious, since it doesn't total to 100. It seems unlikely that all the detail is interesting; consider in particular the "Index map", "Doc. lens", and "Appr. lens" rows, which could presumably be gathered into a single row with a label such as "Other" or discarded altogether. Some of the information might be better in a diagram; for example, a diagram could be used to show what proportion of total time was consumed by each activity.

Results

Task	Size (Mb)	CPU (Hr:Min)	Memory (Mb)
Pass 1:			
Compression	4.2	2:37	25.6
Inversion	6.4	3:02	18.7
Overhead		0:19	2.5
Total	10.6	5:58	46.8
Pass 2:			
Compression	607.9	3:27	25.6
Inversion	137.8	5:25	162.1
Overhead		0:23	2.5
Total	745.8	9:15	190.2
Overall	756.4	15:13	190.2

The overall size of compressed index and text is 36.8% of the size of the indexed data.

A revision of the table on page 157. The percentage column has been replaced by a single line of explanatory text. The "Output" column has been deleted; since most of the values in this column are small, they are relatively unimportant and could if necessary be discussed by the speaker. The space created by deleting a column has eliminated the need for contractions.

Exercises

The skill of good writing is acquired through practice. Pushing yourself, deliberately testing your ability to write new kinds of material and to write faster and better, can make a remarkable difference to the ease with which you can create polished text. Below is a series of exercises, intended not just for novice writers but also to help more experienced writers test and maintain their skills.

Some of these exercises are self-contained; others will be most helpful if adapted to your area of research, in particular by involving papers or passages that are relevant to your work. Educators may wish to choose standard papers and passages to be used by their students.

These exercises require substantial effort to complete—don't expect to run through one or two in a few spare minutes. Set aside a block of time that will be free of interruptions, say two hours, and in that time aim to do one exercise well. The exercises are loosely ordered by the kind of activity they involve, so if you only do a few choose them carefully.

1. Choose a paper from your research area and write a brief answer to each of the following questions.

   (a) What are the researchers trying to find out?

   (b) Why is the research important?

   (c) What things were measured?

   (d) What were the results?

   (e) What do the authors conclude and to what do they attribute their findings?

   (f) Can you accept the findings as true? Discuss any failings or shortcomings of the method used to support the findings.

   (These questions are not just an exercise: to some degree you should ask them for every paper you read.)

   Justify your opinions as carefully as you can. As part of the answers to these questions you should summarize the proposed method and the results achieved. The answers should be substantially your own writing, not quotes, paraphrases, or illustrations from the paper.

   Alternatively, use the questions on page 131 to assess the paper.

2. Choose a paper, perhaps the same paper as for Exercise 1, and criticize the structure and presentation.

   (a) Is the ordering reasonable (of sections and within sections)?

   (b) Are sections linked together?

   (c) Does the paper flow? Are important elements appropriately motivated and introduced?

   (d) Where is the survey?

   (e) Is there a non-technical introduction?

   (f) How carefully has the paper been edited?

   (g) Are there aspects of the presentation that could be improved?

   Based on your criticism, write a review of the paper, including a recommendation as to whether to accept or reject. Take care to discuss all of the paper's major problems.

   Now read your review as if you were the paper's author. Is the review fair or harsh?

3. Some journals have special issues of a series of papers on a related topic. Choose two (or more) papers presenting a similar kind of result and compare them. Have the authors designed and organized the papers in the same way? Where the design choices differ, is one of the alternatives preferable?

4. Some of the papers in the *Communications of the ACM* argue for an opinion rather than present technical results; for example, there are often papers on legal or ethical issues or about computing practice. Choose such a paper and answer the questions in Exercise 1. Carefully analyze the argument used to defend the author's opinion, identifying the major steps in the reasoning. Are the conclusions sufficiently justified?

5. Choose a paper with substantial technical content from *ACM Computing Surveys*. In many such papers the authors are placing their own work in the context of other research results in the area. Do you regard the survey as fair? That is, is the survey an unbiased reflection of the relative strengths of the work in the area?

6. Choose a journal paper presenting new technical results. (Journal papers are usually more carefully written and revised than are conference papers.) Based on the content of the introduction—you should not read the rest of the paper—do the following tasks.

   (a) Identify the hypothesis.

   (b) Suggest a suitable methodology for testing the hypothesis.

   (c) Suggest an organization for the paper, with headings and specific suggestions for the content of each section.

   Now compare your proposals to the body of the paper. Where there are differences, decide which alternative is better. The authors had much more time to think about the paper than you, but are there any problems with the original organization?

7. Summarize a passage, perhaps the introduction of a paper, by jotting down the important points. These notes should be as brief as possible. Now write your own version of the passage using only your notes, without reference to the original. (Mary-Claire van Leunen attributes this exercise to Benjamin Franklin [17].)

8. Choose a popular article about computer science (from Scientific American, say) and summarize it in 500 words. Put the summary aside for a day or two, then review it. Did you include all the important details? Have you represented the article fairly? Would a reader of the summary arrive at the same conclusions about the work as a reader of the original article?

9. Iteratively edit a passage to reduce its length. Start with a passage of, say 300 words, then reduce it in length by 10%, that is, about 30 words; then reduce by a further 30 words; and so on, for at least seven iterations. (To make this exercise more challenging reduce by *exactly* 30 words at each step.) The aim at each step is to preserve the information content but not necessarily the original wording.

   Consider the resulting sequence of versions. With this exercise it is not uncommon for the passage to improve during the first couple of iterations, then become cryptic or incomplete as the text becomes too short for the content. Rate the versions: Which is best? Which is worst?

10. Rewrite the following passage to make it easier to understand. You may find it helpful to introduce mathematical symbols.

    > The cross-reference algorithm has two data structures: an array of documents, each of which is a linked list of words; and a binary tree of distinct words, each node of which contains a linked list of pointers to documents. When a document is added its linked list of words is traversed, and for each word in the list a pointer to the document is added to the word's linked list of documents. An order-one expansion of a document is achieved by pooling the linked lists of document pointers for each word in the document's linked list of words.

11. Choose a passage of 1000 words or so, either a piece of your own work or any passage you understand well. Revise it to improve the writing—that is, edit for flow, expression, clarity, and so on. Make the changes on paper, then type up the result, retaining the paper copy as a record.

    Put the revised passage aside for a few days, then repeat the exercise. Aim to make significant further improvements. (Did you undo any of the previous changes?) Revise again after a break of a few days; and continue until you have five revisions in total. Such revision is the best way to learn how to produce really good text, and many of the best writers revise this thoroughly.

12. Revise a mathematical argument to use less mathematics and more explanation. In a paper with a long proof or mathematical argument, identify the pivotal points of the argument. Is the argument complete? Are too many or too few details provided?

13. Choose a familiar algorithm and a standard description of it. Rewrite the algorithm in prosecode. Now choose an algorithm with a complexity analysis. Rewrite the algorithm as literate code, incorporating the important elements of the analysis into the algorithm's description.

14. Design an experiment to compare two well-known algorithms for solving some problem. An elementary example is binary search in an array versus a hash table with separate chaining, but a more sophisticated example such as a comparison of sorting algorithms will make the exercise more interesting.

    (a) What outcome do you expect—that is, what is the hypothesis?
    (b) Will successful results confirm an asymptotic complexity analysis?
    (c) What resources should be measured? How should they be measured?
    (d) What are appropriate sources of test data?
    (e) To what extent are the results likely to be dependent on characteristics or peculiarities of the data?
    (f) What properties would the test data have to have to confound your hypothesis?
    (g) Is quality of implementation likely to affect the results?
    (h) In the light of these issues, do you expect the experiment to yield unambiguous results?

Selected bibliography

Below is a list of books that I found interesting or valuable in the course of writing this book. Most of them have contributed to my own writing as a practising scientist. Not all of the entries are on style; some are about the process of science.

The first book every writer needs is a good dictionary. I use the Collins English dictionary almost exclusively; it gives both British and American spelling and, in contrast to some other dictionaries, for technical terms often gives the meaning appropriate to computing or mathematics. But there are many reasons for choosing a particular dictionary and you should make your own decision. A rule of thumb is that a dictionary small enough to conveniently carry around is unlikely to be satisfactory.

In addition there are several books on style I use with some frequency. If I was only allowed to keep one such book it would be Partridge [6] or Fowler [2]. If I was allowed one more, it would be Gowers [3] or Strunk and White [7]. Of the books on general science writing, the best is probably O'Connor [19].

General writing style

[1] *The Chicago Manual of Style*, Thirteenth edition, University of Chicago Press, 1982.

A code of rules and judgements on every imaginable issue of writing. Revised many times over the last century, it considers almost every important topic of style. The Chicago Manual is not, however, light reading. Its purpose is to define a consistent style and so it offers judgements on every topic, but they are not always warranted by popular usage. A comprehensive, useful book, but not of great value to most writers.

[2] H. W. Fowler, *Modern English Usage*, Second edition, Oxford University Press, 1965.

Arguably the best, and the best-known, guide to English, this is in effect a collection of over a thousand brief essays, organized as a dictionary. Each is on a topic such as usage of a word or a point of style. Fowler was a vocal advocate of the principle that English does not have to be stuffy to be correct and clear; a principle for which this book provides many examples. It not only reflects good English usage, but to some extent has shaped it.

This edition, a revision by Gowers, is now over thirty years old; some of the entries are becoming dated. However, a new edition is now available.

[3] Ernest Gowers, *The Complete Plain Words*, Third edition, Penguin, 1986.

A thorough discussion of how to write clearly, economically, and in simple language. Concerned primarily with writing for the public, this book is more valuable than some of the specialist style guides. Originally written by Gowers, it has been revised first by Bruce Fraser then by Sidney Greenbaum and Janet Whitcut. It represents a union of diverse approaches and is as a consequence much strengthened: it is a pleasure to read and authoritative in its opinions.

[4] Mary-Claire van Leunen, *A Handbook for Scholars*, Knopf, 1985.

A detailed discussion of some aspects of scholarly writing, in particular style for citations, quotations, and bibliographies. An excellent book, but perhaps of limited value for most authors in computer science.

[5] Frank Palmer, *Grammar*, Penguin 1971.

A good nontechnical introduction to modern grammar. The first chapters discuss the shortcomings of traditional or prescriptive grammars for English.

[6] Eric Partridge, *Usage and Abusage*, Penguin 1973.

An alternative to Fowler, it consists of brief discussions of many words and points of style and like Fowler is organized as a dictionary. It is equally well-informed, has similar scope, and is a little more recent. I would judge Partridge's choice of topics to be somewhat more pertinent to a scientist.

[7] William Strunk and E. B. White, *The Elements of Style*, Macmillan, 1979.

An introduction to the principles of style for English text. It is less than a hundred pages long but more useful than guides many times its length. An essential book.

Style for technical writing

[8] David F. Beer (ed.), *Writing and Speaking in the Technology Professions: A Practical Guide*, IEEE Press, 1992.

A miscellany of articles on different aspects of technical writing and oral presentations. Not all of it is valuable but the various viewpoints are interesting.

[9] Gary Blake and Robert W. Bly, *The Elements of Technical Writing*, Macmillan, 1993.

A succinct and accessible guide to technical writing. There is only a little specific guidance for writing of technical articles, but the advice on usage is highly relevant. An alternative to Cooper.

[10] Bruce M. Cooper, *Writing Technical Reports*, Penguin, 1964.

A basic, accessible book on writing technical material, it covers style, organization, choice of material to include, choice and selection of figures, and other topics. It is not particularly intended for science writing, but is relevant nonetheless. I like Cooper's down-to-earth approach—this book is particularly appropriate for people who are uncomfortable with writing.

[11] Robert A. Day, *How to Write and Publish a Scientific Paper*, Third edition, Oryx Press, 1988.

A broad look at the process of writing and publishing. It is comprehensive and informative but strongly focussed on the biological sciences, which, judging from Day, have rather different conventions to computing. A possible alternative to O'Connor.

[12] Anne Eisenberg, *Guide to Technical Editing*, Oxford University Press, 1992.

Written as a guide for editors of technical writing, this unusual book is also an advanced guide to good style. The first half consists of a series of substantial examples, illustrating different kinds of errors and how they might be corrected. There are also exercises and a valuable dictionary. This book is not for everyone, but is of particular value for people who correct the work of others.

[13] Frances B. Emerson, *Technical Writing*, Houghton Mifflin, 1987.

A comprehensive, detailed introduction to technical writing, or, more accurately, to the duties of a technical writer—for example, job hunting and correspondence are covered in addition to report writing. Emerson has included a brief style guide but the best parts of this book are the chapters on definition, description, and argument; I have seen no other book that covers these topics in any detail.

[14] Leonard Gillman, *Writing Mathematics Well*, Mathematics Association of America, 1987.

Although only fifty pages long, this booklet is as informative as some much longer volumes. Gillman is almost entirely concerned with presentation of mathematics—for other topics of style, look elsewhere.

[15] Nicholas J. Higham, *Handbook of Writing for the Mathematical Sciences*, SIAM, 1993.

Perhaps the best available text on writing for mathematics, Higham covers topics such as style, usage, and overheads. It also includes introductions to some current tools such as emacs, TEX, and ftp.

The material on tools is too brief and superficial for computer scientists, and some topics are covered rather lightly, but the material on mathematical style is excellent. I have aimed, in this book, to write a comprehensive introduction to style for computer science; Higham largely achieves this aim for mathematics.

[16] John Kirkman, *Good Style for Scientific and Engineering Writing*, Pitman, 1980.

Primarily concerned with what in this book I have called tone. It is interesting for the arguments in favour of using personal, immediate, concrete tone in scientific writing. Kirkman has included a detailed report on his survey of attitudes to writing styles.

[17] Donald E. Knuth, Tracy Larrabee, and Paul M. Roberts, "Mathematical Writing", Mathematical Association of America, 1989. (Also available as Technical Report STAN-CS-88-1193, Stanford University, 1988.)

Course notes from a 1987 subject on Mathematical Writing offered at Stanford. This report largely consists of summaries of the lectures, but there are also exercises, discussion of the solutions, and lists of points on writing style. There is much valuable information here, weakened somewhat by the volume of less interesting material and the lack of an index. This report was an important influence on my approach to style for mathematical writing.

[18] Carole Mablekos, *Presentations That Work*, IEEE Engineers Guide to Business Series, 1991.

A guide to oral presentation of technical material. This book is rather slight, but it does have useful tips, exercises, and checklists.

[19] Maeve O'Connor, *Writing Successfully in Science*, Chapman & Hall, 1991.

Perhaps the most comprehensive text on science writing. Topics include organization, style, references, presentations, and issues such as authorship. Its main weakness is probably the breadth of disciplines to which it is appropriate—there are many specific issues that O'Connor does not consider. But overall an excellent introduction to the process of writing and publishing a scientific paper.

[20] Edward R. Tufte, *The Visual Display of Quantitative Information*, Graphics Press, 1983.

A comprehensive discussion of the presentation of figures, in particular graphs. For data or ideas that do not seem to lend themselves to pictorial representation, this book could well have a helpful suggestion. There is also extensive consideration of aesthetics. I enjoyed the historical material; as examples, Tufte has used figures dating back to Leonardo da Vinci.

[21] *The Universal Encyclopedia of Mathematics*, Pan, 1976.

Not a style guide, but a compilation of information on mathematical basics. As well as being a handbook of formulas and methods it is indirectly a guide to usage and nomenclature, and provides innumerable examples of well-presented mathematics.

The process of science

[22] Wayne C. Booth, Gregory G. Colomb, and Joseph M. Williams, *The Craft of Research*, University of Chicago Press, 1995.

An introduction to the process of research, from the initial development of a research question to construction of a sound argument. The later part of the book contains a brief style guide and questions that writers can use to analyze and improve their text. Much of this book is of material not covered elsewhere. A valuable read.

[23] Peter B. Medawar, *Advice to a Young Scientist*, Pan, 1981.

A compendium of advice that scientists should give to their graduate students and research assistants. Concise and direct, it considers why science might be appealing as a career as well as topics such as good conduct and the scientific process.

[24] Anthony O'Hear, *An Introduction to the Philosophy of Science*, Oxford University Press, 1990.

A clear, broad introduction to the concepts underpinning research and the process of science. There are many texts that cover this material, but O'Hear is more succinct and accessible than most. Some familiarity with the basis of science and with concepts such as falsification is I believe essential for an active researcher.

[25] E. Bright Wilson, Jr., *An Introduction to Scientific Research*, Dover, 1952.

An older text, but still a classic. Of particular value to computer scientists is the material on scientific method, design of experiments, and statistical analysis of results.

[26] Lewis Wolpert, *The Unnatural Nature of Science*, Faber and Faber, 1992.

A series of linked essays on aspects of research, including serendipity, creativity, the process of science, pseudo-science, and why technical innovation is not science. Together these essays help illustrate the nature of science—a substantial achievement.

Resources on the internet

Some good material on the subject of writing, research, and refereeing is available via the internet. I maintain a list of pointers to such publications in the page

`http://www.cs.rmit.edu.au/~jz/writing.html`

Index